# Radio Orienteering The ARDF Handbook

*by*

**Bob Titterington, G3ORY**
**David Williams, M3WDD**
**&**
**David Deane, G3ZOI**

**Radio Society of Great Britain**

Published by the Radio Society of Great Britain,
Lambda House, Cranborne Road, Potters Bar, Herts. EN6 3JE, UK

First Published 2007



ISBN 9781-9050-8627-6

*Publisher's note*

The opinions expressed in this book are those of the authors and not necessarily those of the RSGB. While the information presented is believed to be correct, the authors, the publisher and their agents cannot accept responsibility for the consequences arising for any inaccuracies or omissions.

Production: Mark Allgar, M1MPA
Cover design: Dorotea Vizer, M3VZR
Subediting: George Brown, MW5ACN
Typography: Chris Danby, G0DWV

Printed in Great Britain by Latimer Trend of Plymouth

**This book has a supporting website**
**http://www.rsgb.org/books/extra/ardf.htm**

In addition to the extra material mentioned later in this book the site contains any corrections and points of clarification that may not have been incorporated into this printing of the book.

# Contents

INTRODUCTION AND HISTORY .......... 1
YOUR FIRST EVENT .......... 9
PLOTTING BEARINGS .......... 29
THE TOP 10 TIPS TO IMPROVE PERFORMANCE .......... 37
ORGANISING AN EVENT .......... 45
TRAINING AND PRACTICE .......... 57
THE INTERNATIONAL PERSPECTIVE .......... 67
EQUIPMENT .......... 75

# Preface

## PREFACE

International Amateur Radio Direction Finding (ARDF), in its current form, was developed in the years immediately following the Second World War. The current International Amateur Radio Union (IARU) format is of timed competitions in the 3.5 and 144MHz amateur bands. These take place entirely on foot. Amateurs across the world enjoy being able to combine their interest in radio with the opportunity to compete in the great outdoors.

The UK was a late developer in the world of International ARDF, the first event using IARU rules was not held here until 2002. Development has been rapid, and a regular series of competitions is now established. UK amateurs involved in that first event still remember the steep learning curve experienced by both organisers and competitors and are thus in an excellent position to guide newcomers to the activity.

The aim in writing this book was not to produce an encyclopædia of ARDF. Three contributors have come together to provide an introduction and guide to topics that will contribute to success in and enjoyment of this radio sport:

*Bob Titterington, G3ORY* is the Chairman of the RSGB ARDF Committee and first travelled to Europe to compete in the mid-1990s. He was involved in the staging of the first event in the UK.

*David Williams, M3WDD* comes from an orienteering background, and contributes his advanced map-reading and navigational skills and also many lessons of how to organise events.

*David Deane, G3ZOI* derives great personal satisfaction from developing new ideas for low-power transmitters and portable receivers. David has designed and constructed a range of modern, robust and simple-to-operate equipment, including both receivers and transmitters that have proved successful in international competition and home events.

CHAPTER 1

# INTRODUCTION AND HISTORY

Direction Finding (DF) first started in the UK in the late 1920s, more or less simultaneously in Sheffield and at the Slade Club in Birmingham. These early events took place on 1.8MHz and there is thus a long tradition of DF on this band in the UK. The 1950 Council Cup was donated as its name suggests by the Council of the RSGB, and has been awarded to the winner of the National Final every year since then with the exception of the war years. Today, top band (1.8MHz) direction finding in the UK is organised by the British Top Band Direction Finding Association which is a Society affiliated to the RSGB.

When the 2m mobile radio arrived in the mid 1970s, to be followed a decade later by the handheld, clubs started running DF hunts on that band. FM was (and usually still is) employed because that was all that was provided on the radios. The presence of the limiter in the FM receiver did nothing to enhance the effectiveness of the radio for direction finding where detection of changes in signal strength as the aerial is turned is the key to success.

Competitions of this type, usually involving the use of motor vehicles, take place in many countries and are often organised as a summer Club activity.

Internationally, things took a rather different route, away from the car and more towards an orienteering style of competition using specially drawn maps of forest, woodland and open areas of land.

Things seem to have started in earnest in about 1947 when early Swedish DF hunts took place on 3.5MHz. There was also a lot of interest from the former Communist countries, especially Yugoslavia as it was then called. It is important to remember that in the immediate post war years, the only countries in Europe to have transmitting licences for 1.8MHz were Great Britain, Eire and Czechoslovakia. Consequently, other European countries wishing to indulge in direction finding using surface wave propagation were forced to choose 3.5MHz. The legacy of this remains to this day and competitions still use 3.5MHz with 144MHz at VHF.

An unofficial European Championship was held at Sarajevo in 1958. The first proper European Championships followed in 1961 in Sweden, when eight countries participated and both 3.5MHz and 144MHz were used for the first time.

The 1978 International Amateur Radio Union (IARU) Region 1 Conference was a major turning point when a permanent Amateur Radio Direction Finding (ARDF) Working Group was established. Most important of all, it brought ARDF under the IARU umbrella and heralded the start of regular international competition. The first 'official' set of rules was approved at the Region 1 general conference in 1984 and this embodied the principles of competition on the two bands, five transmitters operating in sequence on the same frequency and a standardised system of transmitter identification. The rules have been amended several times since 1984 but, for all practical purposes, these changes are just minor tweaking. The bi-annual World Championships commenced in 1980 at Cetniewo in Poland and the 13th Championships were held in Bulgaria in 2006.

In the early 1990s a small number of UK amateurs began attending events on the continent. The Belgian and German Championships were, and still are, popular with UK participants. Also attracting UK participation were the IARU Region 1 Championships, which are held in odd-numbered years, and the World Championships held in even-numbered years. The Region 1 event took over from the European Championships in 1993.

The first event in the UK using the IARU rules took place in 2002 when a group of amateurs from the midlands built a set of 144MHz transmitters complete with the necessary PIC timers. These timers ensured that each one of the five transmitters in the set transmitted in its allocated 60-second slot in a five-minute cycle. The cycle repeated until switched off or the batteries went flat. All the transmitters operated on the same frequency. There was a delay facility which allowed the clocks on all the transmitters to be synchronised and a time delay set. The transmitters then came on the air in their appointed time slots after a delay of between 0.5 and 15.5 hours, adjustable in half-hour steps. The purpose of this was to eliminate cheating, since the transmitters were timed to come on the air only after all the competitors had deposited their receivers in a supervised 'pound'. Hence, it was not possible to take furtive bearings of the hidden transmitters (or 'foxes' as they are sometimes known).

ARDF using the internationally-agreed rules had finally arrived in the UK.

**GUIDE TO THE RULES**

The full rules can be read on the Internet. Check out **www.ardf-r1.org/rulesv27b.pdf** (this is a pdf file requiring *Acrobat Reader*).

The purpose of the following section is to explain the rules as straightforwardly as possible for the newcomer.

**Transmitters**

The first thing to appreciate about an IARU-style DF hunt is that there can be as many as six transmitters on the air. Five of these are the 'foxes' or hidden transmitters and these are all on the same frequency and are 'time-division multiplexed'. That is to say, they come on the air one after the other, but they all operate on the same radio frequency, as **Fig 1.1** shows.

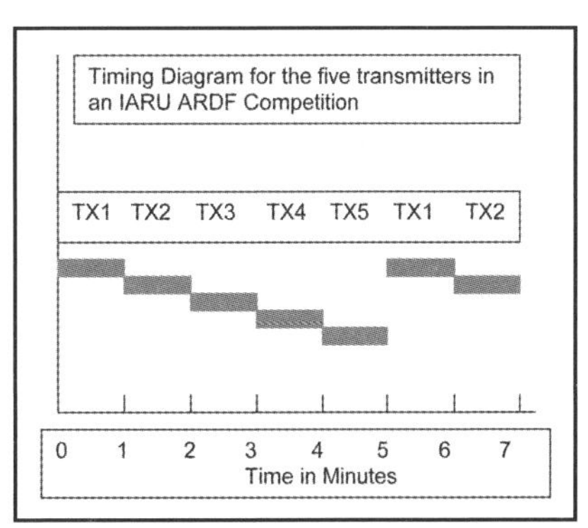


**Fig 1.1: Timing diagram showing how the five hidden transmitters transmit in turn on a single frequency. The competitor does not have to re-tune to hear each of the five transmitters.**

Each transmitter is allocated a one-minute slot in a five-minute cycle. During this time it sends its ID continuously in Morse code, but this is not a problem to anyone who does not know any Morse, since the system has been specifically designed so that a full knowledge of the Morse alphabet is not required. Only the ability to count dots is needed to identify the transmitters.

The transmitters are identified by the letter sequences MOE, MOI, MOS, MOH, and MO5. In Morse, MOE comes out as 'dah dah gap dah dah dah gap dit'. To distinguish them, all you have to do is to count the dots which are sent following all the dashes of the M and the O.

The letters M and O not only form a clear introductory signal to tell the competitor when to start counting the dots, but they also ensure that the transmitter is spending a lot of time radiating a signal rather than pausing between symbols. This makes it easier and quicker to assess the direction of arrival of the signal by swinging the receive antenna from side to side and listening to the variation in signal strength. **Table 1.1** summarises the transmitter identities and the Morse signals radiated by each one.

| Tx | Morse ID | Sounds like | Dots at end |
|---|---|---|---|
| 1 | MOE | Dah, dah, gap, dah, dah, dah, gap, dit | 1 |
| 2 | MOI | Dah, dah, gap, dah, dah, dah, gap, di-dit | 2 |
| 3 | MOS | Dah, dah, gap, dah, dah, dah, gap, di-di-dit | 3 |
| 4 | MOH | Dah, dah, gap, dah, dah, dah, gap, di-di-di-dit | 4 |
| 5 | MO5 | Dah, dah, gap, dah, dah, dah, gap, di-di-di-di-dit | 5 |

**Table 1.1: The transmitter identities and corresponding signals.**

It takes about four seconds to send 'MO5', which is slow enough for all the dots and dashes to be picked out easily.

The licence conditions in the UK and the USA require that the callsign of the 'operator' of the unattended transmitters is radiated at intervals.

**The 'kite' at a transmitter site. This marker [resembling a plastic bag in this black and white photo – *Ed*.] is clearly visible but, in contrast, the transmitter is hard to see. Black and brown materials are used for the external surfaces and, as a result, the transmitter blends into the background.**

Therefore, there is a burst of much higher speed Morse at the beginning or the end of the transmission which is the callsign being sent.

These five transmitters are placed in the competition area which is accessible only on foot, and covered by a specially-produced large-scale map which is issued at the start. The actual transmitters are concealed to some extent, but the site of each is clearly marked with an orienteering style orange and white 'kite' marker (the plastic bag – *Ed*.) to which is attached an orienteering needle punch or an electronic 'punching' device which must be used to prove that the transmitter has been located. The photograph illustrates this.

In international competitions, the transmitters will be 'manned' in the sense that they are under observation. It is usual in these competitions for the competitors to wear race numbers and the transmitter operator will note those that are visible as the competitors arrive at the transmitter.

The sixth transmitter is a beacon transmitter which is on the air continuously on a different frequency. This is located at the start of the finish 'funnel', a couple of hundred metres from the finish line itself. If a competitor gets totally lost, finding the direction to this beacon indicates the direction to the finish. This transmitter radiates the identifier 'MO' in Morse continuously and it operates on a different frequency from the hidden transmitters. In major competitions, competitors are required to punch at the beacon as well as at each of the foxes they are required to find.

Both the frequency of the beacon and the frequency of the foxes are given out before the start and some competitors write them on the receiver as a reminder

**Transmitter Siting**

There are specific rules which govern the siting of the hidden transmitters. They cannot be placed nearer than 750m of the start and 400m of the finish. Also, the transmitters must be at least 400m from each other. In other words, there are areas of the map that are 'off limits' for transmitters and most competitors will mark the 750m zone around the start and the 400m zone around the finish as soon as the map is issued. If the start and finish are co-located, as they usually are in smaller competitions, there is a single 750m zone in which no transmitters are placed.

The map issued at the start has only the start and finish marked. The start is denoted by a red or purple triangle and the finish by a similarly-coloured double circle, ie one circle inside another.

**Classes of Competition**

The existence of several classes makes the sport enjoyable for all ages. In the table below, ages count from January 1st of the year in which you attain the age, eg a man whose 60th birthday is 26 August 2006 is allowed to compete as an M60 from 1 Jan 2006. The exceptions to this rule are the M19 and W19 classes where competitors remain in the class until the end of the year in which they attain 19 years of age.

| Men's Classes | Women's Classes |
|---|---|
| M19 - males aged 19 and under | W19 - females aged 19 and under |
| M21 - males of any age | W21 - females of any age |
| M40 - males aged 40 and over | W35 - females aged 35 and over |
| M50 - males aged 50 and over | W50 - females aged 50 and over |
| M60 - males aged 60 and over | |

**Table 1.2: The classes used in competitions.**

This arrangement means that competitors do not change class in mid-season. **Table 1.2** summarises the age classes as they existed at the end of 2006.
Although the rules allow for flexibility, typically only the M21 class has to find all five transmitters. Classes M19, M40, M50, W19, W21 and W35 have to find four, but the same four transmitters are not allocated to every class. M60 and W50 have to find just three transmitters. More information can be found in Chapter 2.

**Prevention of Cheating**

To stop people taking furtive bearings before they start, all receivers and all competitors are separately corralled before the transmitters come on the air. This rule is applied at all international competitions and major domestic events. At local events the rule is often not applied so that competitors are able to turn up when it suits them rather than everyone having to be present at the very beginning and face the possibility of a long wait in the corral before starting. In this latter case the competitor is trusted to observe the spirit of the rules by not taking bearings before he/she starts.

Competitors are not allowed to communicate with anyone outside the corral and, understandably, international competitions forbid the use of mobile phones in the start corral. The arrangement can lead to a long wait in a big competition since starts are at five-minute intervals.

Competitors are directed to collect their receivers 10 or 15 minutes before their start times. The receivers can be switched on only after the start signal, which is given just as transmitter 1 (MOE) comes on the air and competitors enter a 'start corridor' or marked route leading to the actual competition area. Competitors are allowed to assess the bearing to transmitter 1 as they move down the start corridor, but are not permitted to stop before reaching the end of it. The rules require that when competitors stop beyond the end of the start corridor to take bearings on all five transmitters, they are out of sight of subsequent starters.

It is the responsibility of each competitor to provide weather protection for his/her receiver while it is kept in the corral, although waterproof sheets and even overhead cover is often provided by organisers in major competitions.

**Dress and Safety**
There are no specific rules about dress. Some people wear orienteering clothing but, for the majority, a variety of dress is seen. Basically, anything you are comfortable in for walking (mostly) and occasionally jogging in the woods is acceptable.

As an aid to personal safety, competitors are encouraged to take a whistle and to remember that the emergency signal is six blasts followed by a one-minute interval then six more blasts and so on.

**Time Limit**
There is a time limit to the competition, usually 90 minutes or 120 minutes. If a competitor finds all the transmitters but is just one second over the time limit the he or she will be placed below anyone who is within the time limit but who has found only one transmitter. Pretty galling! So it is important to have a watch and use it to ensure that the time limit is not exceeded. It is a hard decision to make when deciding to 'cut and run' for home. The time limit is announced before the start.

**Information for Competitors**
The organisers need to pass quite a lot of information to each competitor, for example, the frequencies of the beacon and the foxes, the numbers of the transmitters to be hunted by all classes except M21 (who have to hunt them all), the time limit, and any variations to the IARU rules. Most often a relaxation of the 750m zone around the start which should be free of transmitters. This is often reduced to 400m in small competitions in the UK.

It is the custom in the UK to disseminate this information by giving each competitor an information sheet at registration, and a sample is reproduced in **Fig 1.2**.

**FOXORING**
There is a variant of the sport, foxOring, not covered by the IARU rules, which is popular on the Continent of Europe. It is more akin to Orienteering than to IARU ARDF, and in foxOring competitions the competitor is given a map with control circles marked on it. Close to each circle is located a very low-power transmitter which has to be located. It is necessary to use orienteering skills and techniques to

RSGB and LEI OC – IARU ARDF Competition Burbage Common 18 Sep 2005

## Final Instructions

## Frequencies and Callsigns

| | |
|---|---|
| Hidden stations – 144 MHz | 144.525 MHz |
| Beacon – 144 MHz | 144.125 MHz |
| Hidden Stations – 3.5 MHz | 3.580 MHz |
| Beacon – 3.5 MHz | 3.542 MHz |

## Transmitters to find

Competitors hunt transmitters according to age group:

| | | | |
|---|---|---|---|
| M21 | all five | W21 | 1234 |
| M40 | 1235 | W35 | 1245 |
| M50 | 2345 | W50 | 235 |
| M60 | 134 | | |

## Time Allowed

Ninety minutes for both competitions

## The Beacon

The beacon MO is located at the finish and transmits continuously. The transmissions from the beacon will probably prevent you from hearing the hidden transmitters while you remain near to the start/finish area.

## Results

The results will be posted on the website (http://www.ardf.btinternet.co.uk/index.html) tomorrow.

## Miscellaneous Points

- North lines on the map are magnetic north
- Transmitters are at least 400m apart (as per IARU rules). However, one or more may be less than 750m (but will be at least 400m) from the start/finish
- The map is 200x248mm and will fit into an A4 plastic wallet. **Scale 1:10,000**

## Punching at the TXs

Non-electronic punching is being used today. At each TX there is a needle punch next to the orange and white marker. Use this to punch your control card in the appropriate box on the card.
At TX #2, punch in box 2 etc
Each punch carries a different needle pattern

## Timetable

| | |
|---|---|
| 0930 | Registration opens |
| 1015 | 144 MHz hidden TXs come on the air |
| 1030 | First 144 MHz start |
| 1130 | Last 144 MHz start |
| 1315 | 3.5 MHz hidden TXs come on the air |
| 1330 | First 3.5 MHz start |
| 1430 | Last 3.5 MHz start |
| 1600 | Courses close |

## Cheating

You are not required to deposit your receiver before the start of today's competition. This allows competitors to travel from afar and not have to be here for the corralling of receivers. Please respect this concession by not using your receiver to take bearings on the hidden transmitters before you start.

## Call up

You will be 'called up' five minutes before your start time and issued with a map. The map has only the start triangle and the finish circle marked on it.

## Carry a Whistle!

The emergency signal is 6 blasts repeated once every minute

**Fig 1.2: Sample information sheet for a local event in the UK.**

navigate to the area of the circle. Once there, the transmitter will become audible and can be located using an ARDF receiver.

The sequence in which the transmitters are visited has to be decided by the competitor but the decision is much easier than mainstream ARDF because the approximate locations of all the transmitters are plotted on the map. The event is a radio version of an orienteering score event where different controls attract different points and there is a time limit, often too short to allow all the controls to be visited.

There is no difference in the equipment needed by the competitor, but the transmitters are extremely low power, usually based around a quad gate CMOS chip. Since the transmitters have such a limited coverage and cannot be heard more than 100m or so from the control circle, there is no need for them to be individually identified and usually some form of simple tone modulation or keying is used by way of a distinctive sound to listen for.

CHAPTER 2

# YOUR FIRST EVENT

This chapter is intended as a step-by-step guide to the experience of participating in an ARDF event for the first time. As an aid to easy reference, the sections are numbered as follows:


1 Finding out about the event
2 Finding the venue
3 ARDF registration
3.1 Which class?
4 Before you start
4.1 Footwear and clothing
4.2 Other equipment
4.3 Safety
5 Assembly
6 Pre-start
7 The start line and start funnels
8 In the forest
8.1 Etiquette
8.2 The map
8.3 Navigation – relevant orienteering techniques
8.4 Navigation – common mistakes and how to avoid them
8.5 Finding transmitters – the mechanics
8.4.1 Listening
8.4.2 The banner
8.4.3 Proof of visit
9 The beacon, the finish corridor and finish
10 Results


Radio direction-finding events can range from a friendly transmitter hunt around a country park right up to 'serious' national and international ARDF championships. As 'recreational' courses are often offered at 'serious' events, this chapter will cover the systems, equipment and arrangements that you may typically come across at both formal and informal events.

**1 FINDING OUT ABOUT THE EVENT**
Here 'the web' will be the source of information. Local radio (and orienteering) clubs list details of forthcoming ARDF events on their websites. National societies host fixtures lists. Individual enthusiasts maintain links to national and international sites.

Web addresses can be ephemeral but here are some starters (2007)

| | |
|---|---|
| UK | www.ardf.btinternet.co.uk/ |
| USA | www.homingin.com |
| Australia | www.ardf.org.au/ |
| Germany | www.darc.de/ardf/english.htm |
| Netherlands | www.rdf.demon.nl |

In general, you do not need to be a member of an orienteering club or radio society to compete in informal and regional events. The websites will normally give the date, class of event, frequency band(s) to be used, registration times and travel directions.

Normally, equipment can be borrowed from the organisers by prior arrangement. It is not necessary to have your own equipment in order to go out and try ARDF for the first time.

**2 FINDING THE VENUE**

As you approach the given location, look out for signs. This may say 'ARDF' or 'Radio-O', but if the race is being co-hosted with an orienteering event look out signs incorporating the word 'Orienteering'; often a symbolic representation of the 'kite' or 'banner' used to mark control point locations

**ARDF registration: on arrival, look for the ARDF registration point. In this photograph Alan, G8FMH, checks in competitors for an event.**

On arrival, there may be a marshal who will direct you to the parking area – typically some hard standing, a grassy field (beware low ground-clearance vehicles) or forest tracks. The marshal may hand you a sheet of 'final details' and a registration slip for the 'O' event. If required by the landowner, a parking fee may be collected.

**3 ARDF REGISTRATION**

Having parked up make your way to the ARDF registration – this will be separate from the orienteering registration point (see the photograph). Here there will be people to welcome you, answer questions and collect entry fees.

At registration you will find out

- Details of the map: scale, size, date of survey
- Standard rules, distance from registration to start, arrangements for transfer of clothing from start to finish
- Transmitter frequencies and transmitters to be hunted, by age class
- Competition time limits and course closing time
- Method for 'proving' visit to transmitter – 'control card' or 'chip'

These details are often included on an information sheet (see Chapter 1 Fig 1.2 for an example). Feel free to ask about any of the details that are not clear to you.

Your details will be registered and a start time allocated.
A fee is normally payable at this point. It is usually in the £2 - £5 range.
A control card will be issued or your (hired or own) chip number will be recorded.
At informal events, the map may be issued at registration but, in major competitions, it is issued at the start.

### 3.1 Which class?

ARDF classes are determined in the same way as orienteering classes – but the full range of orienteering classes is not used. M and H stand for men's classes and W and D stand for women's classes. Competitors can 'run' up but not down.

| Class | Transmitters to be hunted* | Age** |
|---|---|---|
| M18 | 3 TXs*** | Under 19 |
| M19 | 1,2,4,5 | Under 21 |
| M21 | 1,2,3,4,5 | 21 and over |
| M40 | 1,2,3,4 | 40 and over |
| M50 | 1,3,4,5 | 50 and over |
| M60 | 3 TXs*** | 60 and over |
| W18 | 3 TXs*** | Under 19 |
| W19 | 1,3,4,5 | Under 21 |
| W21 | 1,2,3,5 | 21 and over |
| W35 | 2,3,4,5 | 35 and over |
| W50 | 3 TXs*** | 50 and over |

* Typically, but check before you start
** In the year of the competition
*** Announced on the day of the competition

**Table 2.1: Typical ARDF classes.**

The table above has an M18 class added to the classes detailed in the IARU rules (see Chapter 1). Making a three-transmitter course available to young people provides a good introduction to the sport of ARDF.

## 4 BEFORE STARTING

### 4.1 Footwear and clothing

No specialist footwear or clothing is required for ARDF, but wearing suitable shoes and comfortable clothes will make the experience more enjoyable.

You will not know exactly what terrain you will be crossing, but you should be prepared for it to be uneven, muddy, potentially slippery and with steep up and down gradients. You may need to cross ditches

**Typical orienteering shoes with rubber studs.**

and ford streams. You may also encounter brambles, fallen branches, exposed rock, sharp stones and other obstacles.

The *ideal footwear* would be specialist orienteering shoes – these bear rubber studs (and optionally spikes) and are constructed of strong water- and tear-resistant materials, as shown in the photograph. These will provide suitable protection and excellent grip – reducing the risk of injury and increasing speed and confidence over rough terrain. Orienteering shoes can be obtained from a number of on-line retailers and at major orienteering events.

**Competitors finishing in the 2006 World Championships, dressed in the type of clothing that is widely used.**

Acceptable alternatives include walking boots, fell running shoes and robust, preferably old, trainers. Tennis shoes and 'racing flats' generally will not afford sufficient grip. Beware trainers with built up heels as there will be an increased risk of turning an ankle.

**Suitable clothing** will provide protection and ensure adequate comfort. Exactly what is suitable for you depends on how fast you intend or wish to travel. If you will be walking then no specialist clothing is required – though it would be a good idea to wear older clothes that you do not mind getting dirty and, perhaps, torn.

**Competitor wearing gaiters to provide lower-leg protection against nettles and brambles.**

The serious competitors will be seen wearing specialist and lightweight orienteering trousers, lycra tights or 'tracksters' and on the upper half of the body combining a lightweight shell layer over a 'wicking' thermal top. Lower leg protection will be in the form of padded gaiters or 'Bramble Bashers' (calf-length socks with a tough plastic layer bonded to the front).

Bear in mind the predicted or potential weather – the organiser may *require* you to carry (or wear) a waterproof rain jacket or cagoule. In cold weather you may wish to wear or carry a woolly hat and gloves – for use in case of injury and to avoid hypothermia. In hot weather, hats and sunscreen may be compulsory.

Note that shorts are hardly ever a good idea and, if the race is co-hosted with an orienteering event, the rules of orienteering *require* full leg cover.

**4.2 Other equipment**

The other equipment you may wish to carry depends on personal choice but here is a checklist of suggestions . You should plan to pocket or fix as much of the equipment to your person as possible – keeping your hands free to carry your receiver and as required at the time – compass, pen or map. See **Table 2.2** for suggestions.

**WHAT TO CARRY**

| Item | Notes |
|---|---|
| **Receiver** | Check that you have the receiver for the correct frequency band (3.5MHz or 144MHz) and that you have made a note of the transmitter and beacon frequencies. There may be an opportunity to 'tune in' to demonstration transmitters before the race. |
| **Battery for receiver** | Check the voltage, preferably under load and/or carry a spare. The recommendation is at least 9V at 25mA assuming that your receiver uses a 9V PP3 battery |
| **Aerial** | On 3.5MHz, the aerial is always integrated into the receiver in the form of a ferrite rod or a loop antenna. On 144MHz, this is not necessarily the case. Because there is so much to carry, it pays huge dividends to have a small lightweight receiver permanently attached to the boom of the 144MHz Yagi antenna. |
| **Headphones** | It is advisable to carry a second pair in case your main pair becomes damaged – the Walkman bud-type will easily fit in a pocket. |
| **Compass & protractor** | Either a hand held, 'thumb', map-board-mounted or on-receiver compass. A hand held compass with a rectangular baseplate can also be used for plotting the bearings – otherwise you will need to carry a protractor. Read Chapter 3 before deciding on your compass strategy. Wearing the compass on a cord round the neck is NOT recommended. |
| **Marker(s)** | A means of marking bearings on the map and a method for attaching markers to your map, receiver or person. Waterproof 'indelible' markers and 'biros' are popular – but beware of obscuring map detail that you may need later. Chinagraph (grease) pencils may be used as these are not affected by water and marks can be erased (as necessary or by accident). At least one international competitor swears by eye-liner pencil – but note that these will melt above 35°C! |
| **Templates** | These are for marking the (typically) 750m and 400m exclusion zones around the start, finish and transmitters. Check you have the correct size for the advertised map scale. See Chapter 4 for more details. |
| **Map 'protection'** | Your preferred means of protecting the map from water, mud and damage. Typically a transparent and heavy-gauge plastic bag, transparent sticky-backed plastic or a rigid sheet of clear plastic. There are several different ways of carrying the map – the simplest is 'by hand' – but this can cause problems as you also have to carry a receiver, compass, control card/chip and pen. Many |

| | |
|---|---|
| | competitors mount their map on a pivoted board and strap this to one arm. |
| **Map** | If this has been issued to you at registration, make sure you have marked any out-of-bounds areas and late changes (map corrections). |
| **Control card or chip** | Control cards may require protection and are often 'safety pinned' to your sleeve or the front of your running top. 'Chips' or 'dibbers' are worn on the finger, held in place by a loop of elastic. As chips cost between £15 and £40, competitors often tie a loop of string round the wrist and attach this to the chip as a backup. |
| **Whistle** | If these are required then a plastic 'survival' whistle safety-pinned in a pocket is best. |
| **Watch** | A waterproof digital watch, preferably with a countdown timer and alarm, is useful. The watch should be synchronised with 'race time' . By watching the seconds you can make sure you get that last accurate bearing before a transmitter goes off the air. By watching the minutes you can make sure you are in a good position for the instant at which the wanted transmitter 'fires up'. |
| **Information** | If you think that you may forget important information (or become confused in the heat of competition), make a written note of the identities of the transmitters you are hunting, the time limit, the course closing time and any safety bearing. |
| **WHAT *NOT* TO CARRY** | |
| **GPS** | In formal competition, the carrying of global positioning satellite (GPS) receivers is *absolutely forbidden*. Although they would be of little use in finding the hidden transmitters and in navigation - there are no grids on orienteering maps - they would be of use in fixing locations which could then be communicated to team mates. |
| **Transmitters** | In formal competition, the carrying of mobile phones and radio transmitters is *absolutely forbidden*. There have been instances of early starting competitors 'phoning back' the location of transmitters to team mates. Teams have been threatened with disqualification when a bag search in the assembly area has turned up an unauthorised Nokia! The rules on GPS and mobile phones are generally not enforced for novices and in informal races - but it would be best to check with the organiser first. |

**Table 2.2: Suggestions for equipment.**

### 4.3 Safety

You will be checked out into the forest and checked back in. It is important to make sure that you pass through the finish funnel and hand in your card or chip - retiring and returning to registration or going straight to your car can lead to unnecessary searches being mounted. If travelling alone, you may be required to leave an item such as a car key at registration - for collection upon your return. In public areas your car registration number may be recorded to assist in checking that everyone is out of the forest.

Staying out after your competition time limit has expired 'to get that last transmitter' is not considered 'good form' and in any event you should return before the course closes so that the organiser can set about collecting in the transmitters and packing up.

It is sensible (and may be a *requirement*) to carry a whistle and to acquaint yourself with the international distress signal (see Chapter 1). Whistles should only be blown in case of injury, hypothermia or other serious distress – as all competitors will be expected to abandon their run and offer assistance – being 'a bit lost' is not considered sufficient cause!

The organiser may indicate a safety bearing: "Head south-west until you reach main road and then turn south".

Beware of hazards such as crags, rock faces, deep pits, rabbit holes, greasy wooden bridges, steep slopes, fences, bogs and marshes – often these will be indicated on the map, but their visibility and 'crossability' is not guaranteed. Exceptional known hazards may be marked with coloured tape. Take note of any specific safety briefings given or danger areas shown on the map.

Specially for ARDF – beware crossing and running along roads and forest tracks – car drivers and cyclists will not be aware that, with headphones on and Morse beeping loudly, you will *definitely not* hear their approach.

A *very few* orienteers have suffered from tick-borne infections - Lyme Disease is now endemic throughout the UK and Europe and Tick-Borne Encephalitis is a risk in central and eastern Europe. You may wish to take precautions (clothing and insect repellent) and then check for ticks after competing.

**5 ASSEMBLY**

At major events, the competitors will be led, as a single group, some way to an assembly area. The departure time for this march will be displayed at registration. At the assembly area, all receivers must be deposited in the marshalled 'box' before the time at which the hidden transmitters are due come on the air. There will be a marked clothing dump where coats, hats and other clothing not required during the race can be left for transfer to the finish.

Starting time lists will be displayed – these show the start time for each numbered start group and list the names of competitors in each start group. There will be a reminder of the frequencies being used and the transmitters to be hunted (by class).

From assembly, competitors are called up to pre-start.

At informal events, the assembly, pre-start and start will normally be combined as a single taped box and will be located close to ARDF registration. Receivers are not collected – competitors being trusted not to listen to transmitters or take bearings before their official starting time.

**6 PRE-START**

**Minus 10**

The start group is called up 10 minutes before the actual starting time. Names are checked off against the starting list, safety checks are performed and any resetting of chips carried out; competitors then follow tapes to the minus 10 line. The -10 and -5 lines are normally positioned so that competitors in the assembly area cannot see the directions in which starters run off.

**Minus 5**

Here, the map is issued. It may already have the start and finish locations marked or there may be a 'master map' from which these may be copied. The start location is indicated by a triangle and the finish by a pair of concentric circles. Note that the *triangle* shows the position of the starting line and not the far end of the starting funnel. The *circles* show the position of the beacon transmitter and not the finish line. Unless otherwise stated, no transmitters will be placed within 750m of the start or 400m of the beacon. Competitors may mark their map with the 750 and 400m exclusion zones around the start and finish (see Chapter 4 for more details).

**7 THE START LINE AND START FUNNELS**

If you have not done so already, synchronise your watch with the official race time – this will allow you to know exactly when each transmitter will start and stop sending, so you can pause and listen carefully to quiet and distant transmissions, position yourself ready, and make sure that that last bearing is truly accurate.

Groups will start at exactly 0, 5, 10, 15, 20... minutes past the hour – meaning that every competitor will hear transmitter number 1 first.

There will be one or more taped start funnels leading away from the start line – each class will be allocated to one or other of these funnels ensuring that competitors are spread out as they enter the forest. At the given start signal you may cross over the start line, turn your receiver on and listen to transmitter number 1. As you pass down your funnel you may take bearings on the transmitters but you should keep moving and not stop at all until you reach the end.

## 8 IN THE FOREST

### 8.1 Etiquette

There may be other users of the area present and ARDF events take place on sufferance – always respect and give way to members of the public – especially as these people may be local and complaints will often get back to the landowner. Beware particularly of horse riders – a runner, head down and headphones on suddenly appearing out of the bushes or charging past will surprise and frighten the steadiest of horses.

In serious competition it is good practice to remain silent – this way you will not disturb the concentration of other runners – nor lead them into the transmitter. Don't offer or invite assistance or indicate the way to the transmitter – unless, of course, the other competitor is a distressed or confused junior. It is best to move smartly away from the transmitter location as soon as you have 'punched' – standing around looking at your map will advertise the location to other competitors.

### 8.2 The map

In theory, and given sufficient time, it would be possible to find all the hidden transmitters without reference to a map. In practice the map will allow you to select suitable routes that maximise the information gained from transmitter bearings, to estimate the reliability of these bearings and to travel through the terrain fast and efficiently – avoiding unnecessary climbing, descent, obstacles and hazards.

Normally an orienteering map will have been provided. This section describes the important features of 'O' maps, with respect to ARDF.

An orienteering map is a topographic map but is not like any map with which you may be familiar, such as those produced by Ordnance Survey, USGS or other national mapping agency. The map will have been specially surveyed and drawn for the purpose of orienteering using special symbols and colour codes. This means that some things you may be expecting will be not be shown (latitude/longitude grid lines, numbers on contours and place names) and specific orienteering information will be shown (vegetation type, small pits, fodder racks, earth walls and even individual boulders). Much of this fine detail shown on the map will be of little use for ARDF! Orienteering maps are internationally standardised – though there may be local specialities – there are not many termite mounds in Herefordshire, but then few apple orchards in central Australia. A legend will be provided that identifies these 'special' symbols. The full International Specification for Orienteering Maps (ISOM 2000) can be found at: **http://lazarus.elte.hu/tajfutas/isom2000/index.html**

The map will printed at a scale of 1:10000 or 1:15000, meaning that one centimetre on the map represents just 100 or 150 metres on the ground. The map will also show the direction of local magnetic north.

The terrain is shown on the map through colours and symbols. The illustrations in this book are black and white only and cannot show a specimen map or a legend. The colours used are explained below.

**Brown - the landform**
Contours, embankments, dry ditches, erosion gullies, pits and excavations are shown in *brown*. Correct interpretation of the shape of the land will allow you to select suitable locations for taking bearings, to judge whether these bearings may be subject to multipath reflections and to select routes that avoid wasting energy through unnecessary climbing.

**Black – man-made features**
Tracks, paths, bridges, walls, fences and power lines are all shown in *black*. Following such features is normally much faster (and more reliable) than cutting directly through terrain. Taking careful note of the location of gates in high fences or bridges over rivers can save a lot of wasted effort.

Beware of metal fences and power lines when taking bearings as these can distort the signal received.

**Black – rock**
Black is also used to indicate cliffs, rock faces and crags. These can be serious hazards.

**Blue – water**
Lakes, rivers, streams, bogs and marshes are shown in *blue*. These may represent serious impediments to progress. Please note that the 'cross-ability' of these features is shown for typical meteorological conditions only!

Beware that areas of standing water can affect signal propagation and the reliability of transmitter bearings.

**White, green and yellow - vegetation**
Confusingly for newcomers, freely runnable forest or wooded land is shown as **white** – this is for historical and ink-saving purposes. Orienteering started in Scandinavia where the terrain is largely forest and mostly easy-going – little bracken or bramble.

However, in less-blessed parts of the world, *green* is used to show decreasing penetrability, the darker the green the harder will be the going - from 'slow running' through to 'impassable'. Normally, in thicker forest, it will difficult to spot the transmitter banner. A green line screen over white is used to show areas having good visibility but with difficult going because of (often seasonal) ground vegetation.

**Yellow** is used to show varying degrees of open land – from scattered trees through rough heath or moorland to fast grassland and standing crops of several varieties. Note that fields may be in or out of bounds – check the map and the information handed out at registration.

**Out-of-bounds and other restrictions**
Permanently out-of-bounds areas may be indicated with an *olive* screen (settlements and private dwellings) or parallel thick black lines. It is important that such areas are not entered or crossed.

Temporary restrictions are indicated with a *purple* overprint – this covers out-of-bounds, fences not to be crossed, routes that *must be* followed and routes that *must not be* followed. Dangerous areas are shown with cross-hatching and there are symbols for crossing points, drinks and first aid stations.

Failure to note these restrictions may result in disqualification and, in addition, risks the loss of the area for ARDF and orienteering.

**Map corrections**
There may have been changes since the map was surveyed and printed. There may be seasonal or temporary restrictions. This information will be marked up on maps which will be displayed in the pre-start and start area. You may be required to mark this extra information on your own map.

## 8.3 Navigation - relevant orienteering techniques

It would be possible to find all the hidden transmitters solely through the use of radio bearings. Understanding and using selected orienteering techniques and being aware of commonly-made navigational errors will help your complete your course *within-the-time-limit*!

**Orientate and 'thumb' the map**
It will be much easier if you keep the map orientated to magnetic north – so as you pass features they come up 'on the correct side' and you will not have to rotate what-you-are-seeing when it comes to choosing the correct path to take at a junction.

Orienteers often hold the map with their thumb pinpointing their 'current position' – making quick glances at the map easier and more reliable.

**Simplify – catching features and handrails**
It is not necessary to read the map all the time – if you are running along a path, or across terrain, identify an obvious feature on the map and run until you see or cross this. Good *catching features* are streams, paths and fences: "I should run down the path until I cross the stream", "If I hit the fence I have gone way too far!"

If you are travelling off-path between transmitters, there is no need to waste time on fine navigation if you can find and use a good *handrail* – these are normally line features like fences, streams and vegetation changes: "Keep the stream on the left", "Run along the edge of the felled area", "Stay on top of the ridge".

**Rough compass**
When travelling across terrain 'on a bearing' do not try to follow the compass all the time – it is faster and more accurate to 'sight' on a tree or other feature a hundred or so metres ahead and then run to this before looking at your compass and repeating the procedure.

**Knowing where you are – 'ticking off'**
It will not be necessary to know your exact location at all times, but maintaining awareness while chasing down transmitters and moving

through the terrain will help a great deal when plotting bearings and selecting routes.

The best way to do this is mentally to 'tick off' features as you pass them: "There is that path junction", "I should see a tower soon", "I have crossed the ditch" and to anticipate "I should be crossing a road soon", "Where is that lake – have I drifted off route?".

Be prepared to stop if things you are seeing are no longer fitting the map or things you are expecting are not in evidence.

**When you don't know where you are – 'relocation'**
It is very likely that while 'chasing down a transmitter', you will lose track of your position and will need to *relocate*. Relocation is a skill that can be learned – here are some guidelines.

- Once you become aware that you are lost – stop and re-orientate the map to north.
- Try to remember where you were when last *certain* of your position.
- Think how far you might have come from this point.
- Look around – what can you see? Are there any obvious features that you can find on the map?
- Think what type of error you might have made (see below).

Orienteers who become totally lost can attempt to retrace their steps or set a *safety bearing* towards a prominent line feature or the forest boundary. In ARDF we have the added assistance of the hidden transmitters and the beacon.

**Route choice**
It is rarely a good idea to run straight through the terrain at a transmitter, unless you are very certain that it is close. Paths tend to be quicker, you are less likely to get lost, and it is easier to pinpoint your position when it comes to plotting bearings.

When choosing a route, take into account the terrain types to be crossed, how much climb and descent will be involved, how many opportunities for taking bearings on transmitters and, most importantly, whether you have the skills to execute the route without getting lost or wasting brainpower on navigation instead of concentrating on listening for transmitters.

**8.4 Navigation: common mistakes and how to avoid them**
**The wrong path**
It is not practical to mark all paths on the map. Small paths appear and disappear with the seasons, new paths may appear as a result of

forestry operations, old paths may disappear through disuse. At a path junction, always use your compass to make sure that you select the correct path and then confirm the direction as you travel down the path.

Make sure that the junction you are at is really the junction shown the map – you may have overshot and missed the junction or have found a new unmarked path. You can do this through estimating the distance that you have travelled, looking for adjacent features and confirming the angles at which the paths join or cross.

**180° error**

It is easy to take a bearing correctly and then to plot it on the map 180° out. This is because the map may be 'upside down' when you plot the bearing.

It is also common, mistakenly, to take a bearing using the 'south end' of the needle and then to correctly plot this incorrect bearing.

Often a competitor will run 180° in the wrong direction again – lining up the south end of the needle with the north arrow in the compass baseplate.

To avoid the 180°-error, be aware that it is an easy mistake to make, check and re-check the needle when taking bearings, note the direction of the map 'north-arrows' when plotting bearings and, when you start to follow your bearing, check the features you expect to see from the map against those you are actually passing.

**90° error**

This may occur when paths and tracks are laid out in a grid system. The competitor does not run across terrain as accurately or a straight as planned and reaches a path or a path junction and runs 90° in the wrong direction. This is a very easy mistake to make – and hard to detect in plantation forest where all the paths look the same and half of them go in the same direction.

Always use your compass to confirm the direction of any path before following it too far.

**Parallel error**

An area may contain many similar features – you arrive at a path bend and everything *seems* to fit. Over-confident that you know exactly where you are, you strike off down the path and then suddenly you become

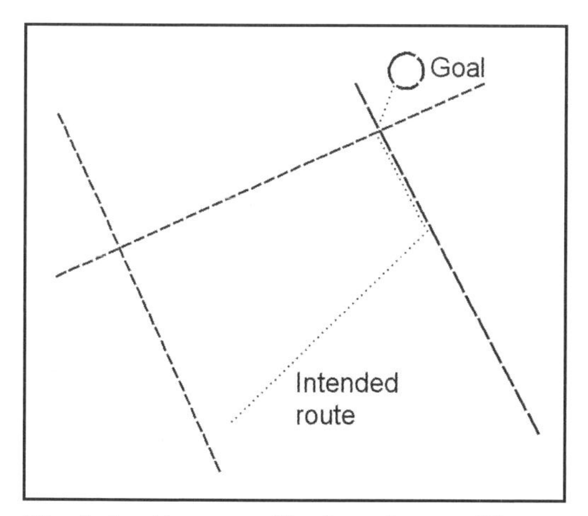


**The intention: run the bearing until you hit the track, turn left and look on the right after the track junction.**

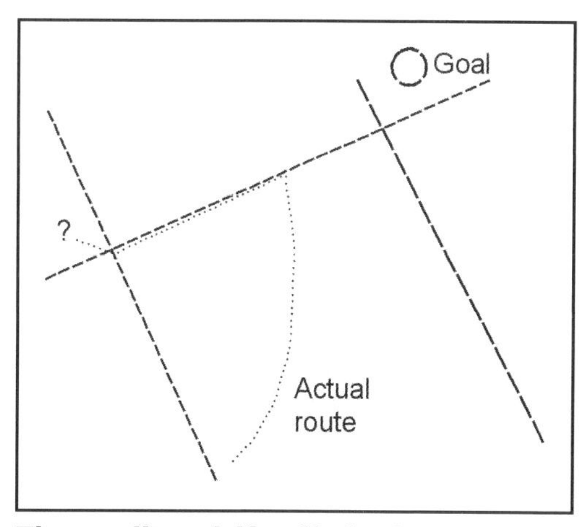


**The reality: drift off the bearing, hit the wrong path, turn left and end up lost, some way from the wanted location.**

aware that you are not at all where you want to be. It turns out that you were on the wrong path all the time.

This is a very easy mistake to make – there is a great temptation, in the face of minor discrepancies, to 'make the ground fit the map'. The way to minimise time lost through parallel errors is always to be aware that you *might* be wrong and to cross-check continually what you are seeing against what you are expecting.

For more information on orienteering techniques, consult: *Orienteering, the Essential Guide to Equipment and Techniques*, by Ian Bratt, ISBN 1-85974-910-0.

**8.5 Finding transmitters – the mechanics**

Techniques and strategies for efficiently locating the hidden transmitters are covered in Chapter 4 – in the current chapter we will deal with the 'mechanics' of identifying which transmitter is currently transmitting, confirming which transmitter you have found and proving your successful visit.

**8.5.1 Listening**

You can visit the hidden transmitters in any order. Each transmitter sounds for just one minute. Number 1 will transmit during the first minute, number 2 during the second and so on until number 5 completes transmitting and the cycle restarts with number 1.

Transmitters identify themselves through Morse code as described in Chapter 1 but summarised below. The end of each transmission may be signalled – by a call sign or a steady tone – use this as your clue to stop and get a 'really good last bearing'.

Distance and intervening terrain may mean that you cannot hear the currently 'live' transmitter – in this case you will have to rely on your

| Tx | Code letters | Code Morse | Minutes* |
|---|---|---|---|
| 1 | MOE | — — . | 0, 5, 10, 15... |
| 2 | MOI | — — .. | 1, 6, 11, 16... |
| 3 | MOS | — — ... | 2, 7, 12, 17... |
| 4 | MOH | — — .... | 3, 8, 13, 18... |
| 5 | MO5 | — — ..... | 4, 9, 14, 19... |
| Beacon | MO | — — | Continuous |

*As displayed on your watch

**Table 2.3: Transmitter identification signals and times.**

watch when positioning to obtain a good bearing as the next transmitter comes on the air.

Periodically check the tuning of your receiver – especially if ALL the transmitters sound weak – it is very easy to knock the tuning knob.

### 8.4.2 The banner

The transmitter and antenna should be hidden but the location will be marked by an orange and white 'banner'. The banner or kite will not be hidden and ideally will be visible from 5 to 10m away from all practical directions of approach. The objective is to site the banner so that competitors have to direction-find to the transmitter and then find the banner without relying on random hunting and good fortune.

**Typical banner and marking of a transmitter site.**

The banner will bear the transmitter number and, if multi-band competitions are taking place in the same area, either '2m' or '80m'. In certain territories, the number may be prefixed with 'FOX' as in 'foxhunting'.

It is good practice not to remain close to the banner for too long – but to 'prove' your visit and depart as rapidly as possible so that your motionless presence will not the give the location away to other competitors.

### 8.4.3 Proof of visit

Proof that you have found the transmitter will be through marking of a control card or recording on an electronic chip. Before using your card or chip make absolutely sure that the banner is marking one of the transmitters that you are seeking and is not part of any co-organised orienteering event!

**Control Card**

This is the traditional method. The competitor fills out the necessary details on both "card" and "stub" before starting: name, callsign, class, start time and club.

The stub will be separated from the card as you start and will be re-united on your arrival at the finish – as part of the safety check. If travelling alone, record your car registration on the stub as this may be used as a backup safety check.

ultrasport
THE ORIENTEERS' SHOP
Nova House,
Audley Avenue Enterprise Park,
Newport, Shropshire. TF10 7DW
Tel: 01952 813918
Fax: 01952 825320

B.O.F. RECOMMENDED DESIGN
NAME
CLASS COURSE
START
CLUB TIME

FOR OFFICIAL USE ONLY
FINISH
START
TIME

| 19 | 20 | 21 | 22 | 23 | 24 | 25 | 26 | 27 |
|---|---|---|---|---|---|---|---|---|
| 10 | 11 | 12 | 13 | 14 | 15 | 16 | 17 | 18 |
| 1 | 2 | 3 | 4 | 5 | 6 | 7 | 8 | 9 |

COMPLETE AND HAND IN THIS SECTION AT THE START
VEHICLE REG NO.
NAME
CLASS
CLUB
COURSE
FINISH
START
TIME

**Typical control card used with non-electronic punching. The competitor uses the pin punch to place a pattern of needle holes in the box on the control card corresponding to the number of the transmitter.**

It is important that, once you have started, you pass through the finish and hand in your card. Even if you find no transmitters the organiser will need to confirm that you are safely out of the forest.

At each banner there will be a plastic 'pin punch' – each punch having a different pattern of pins. Punch in the box on the card that corresponds to the transmitter number. Typically the beacon is treated as being transmitter number 6 and *does* need to be 'punched'.

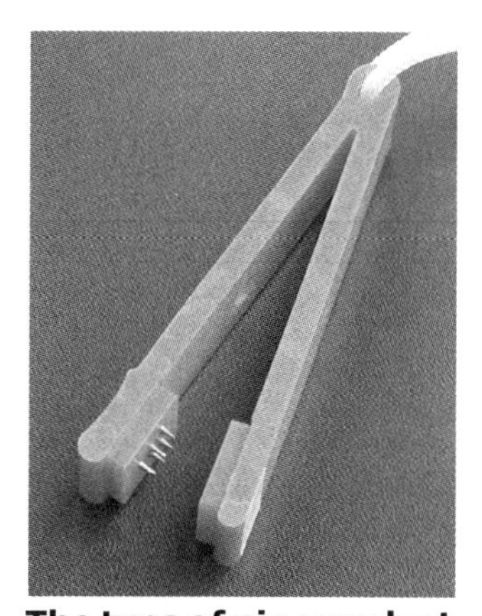

**The type of pin punch at each transmitter. The pattern of the pins is unique and when used to punch a mark in the appropriate box on the control card gives proof that the transmitter has been located.**

**Electronic Chip**

There are two systems in common use - "Sport Ident" and "EMIT" - both of these incorporate a smart-card, which you carry, and RFID "units" which are located at each banner. Your visit is recorded on the smart-card, and optionally on the "unit".

The terminology remains to be standardised and you may find the part that you carry variously described as a 'dibber', 'chip', 'brick' or 'e-card', and the action of recording your visit as 'dibbing' or 'punching'.

Competitors may use their own chip or can hire one for the day, usually at the cost of £1.

**Using SportIdent**

At the pre-start, the card is 'cleared' and then 'checked' in

**The Sport Ident system uses a small box with a hole on the top. A box of this type is located at each transmitter.**

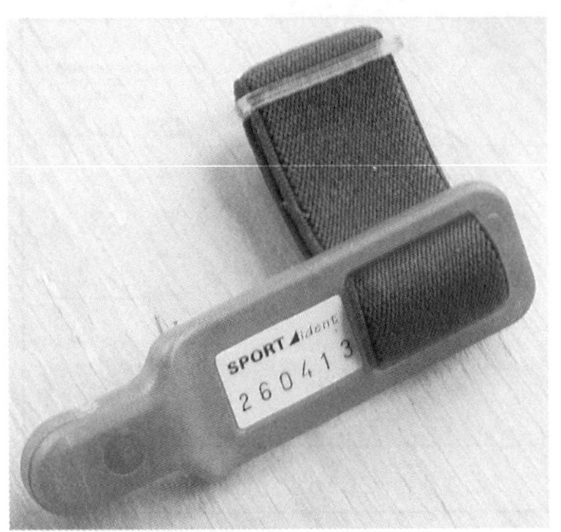


**Photograph of the SI 'dibber' or 'chip' carried by the competitor . This is inserted into the hole in the unit at the transmitter to record on the dibber the time of 'punching' and the identity of the transmitter visited.**

two special units. This deletes any record of previous races and confirms that the card is blank and ready to go.

As you start, it may also be necessary to record your departure time in the start unit (known as a punching start). Alternatively the start punch may be taken a minute before the start time and just used for the safety check.

The SI unit confirms a successful recording by an audible beep and a flashing red LED - you put the narrower end of the chip in the hole and wait for confirmation. If you don't get this - remove the chip from the hole and repeat - if still no luck then look to the backup pin punch and punch your map with this.

**Using EMIT**

Here, the chip is reset, cleared and started by placing it in the start unit. A flashing LED confirms that the start has been recorded. At each banner place the card into the unit, display side up, and 'backup card'-side down, and wait for the LED to flash.

It is not strictly necessary to place the chip all the way into the unit - you can often just wave the chip around close by, but if the unit has failed you will not get the pin mark in the backup card.

You can check that the transmitter has been recorded by looking at the display on the chip.

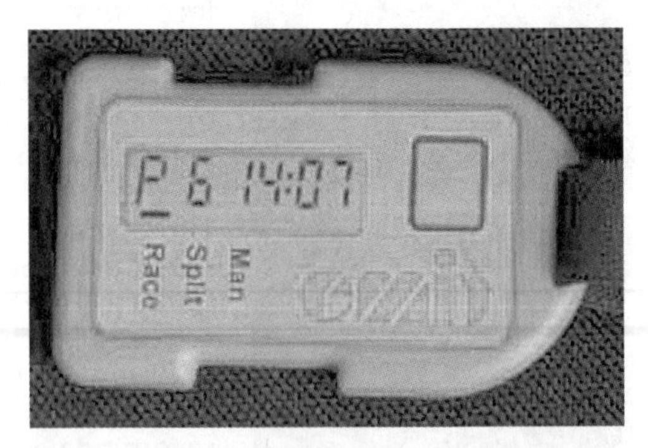


**The EMIT unit carried by competitors. This is version 3 which has a digital display to show timing information at each 'punching'. Each unit has the facility to attach a 'back up' paper card to the rear. When placed on the unit at each transmitter to verify that the transmitter has been found, a mark is made on the card. This can be used to confirm the visit, even if the electronics in the unit are subsequently damaged or erased.**

## 9 THE BEACON, THE FINISH CORRIDOR AND FINISH

When you have found all the required transmitters (or you are running out of time) then head for the beacon. The location of the beacon transmitter will be shown on your map - so you have the choice of orienteering to the beacon - or following a radio bearing - or a combination of these techniques.

Note that the beacon transmits continuously, but on a frequency different from

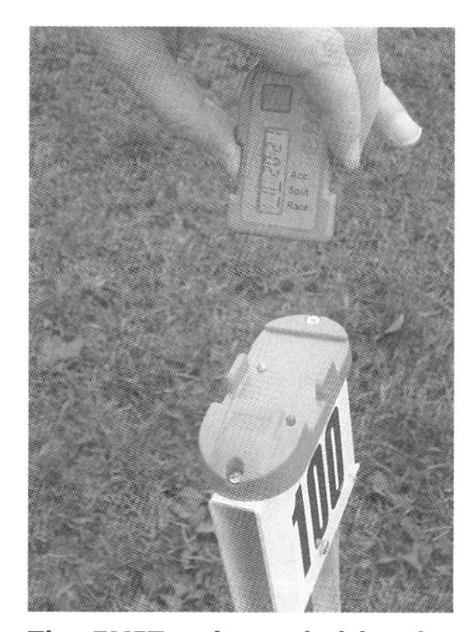


**The EMIT unit carried by the competitor is about to be placed on the unit located with the transmitter to prove that latter has been located.**

that used by the hidden transmitters– you will need to retune your receiver to hear the beacon.

The beacon will be marked with the standard banner, bearing a punch or SI/EMIT unit. It is necessary to 'punch' at the beacon.

At informal events, the beacon may be located at the finish point. If this is also the start point, the beacon may be turned off during minute one to allow competitors to listen to transmitter 1 without interference from the beacon.

At other events there will be a taped corridor or funnel leading from the beacon to the actual finish line. If you come across the funnel *before* you locate the beacon do not enter or run along inside the funnel - you may be disqualified as this counts as impeding other runners. You may be timed across the finish line or there may be a final SI/EMIT unit to record your time.

At the finish, water or other drinks may be provided, you will be reunited with any clothing left at the assembly area and a gentle warm-down will aid recovery.

**10 RESULTS**

Provisional results will be displayed in the finish area. Final results will be posted to the event website.

The winner is the competitor who finds, without timing out, the most required transmitters in the shortest time. Results are ordered by class as follows, a sample being shown in **Table 2.4**.

- For those who finish within the time limit having found all the required transmitters - by shortest running time and then...
- For those who finish within the time limit having found all the required transmitters bar one - by shortest running time and then...
- For those who finish within the time limit having found all the required

transmitters bar two – by shortest running time and then...

- Until all those who finished within the limit, having found at least one transmitter, have been ranked.

Timed-out competitors are not formally ranked but may be shown in order of number of transmitters found and running time.

**Class: M21 Tx: 1,2,3,4,5 Limit: 120 min**

| Position | Name | Tx | Time |
|---|---|---|---|
| 1 | A Andersson | 5 | 51:15 |
| 2 | B Brookes | 5 | 52:15 |
| 3 | C Christian | 5 | 119:20 |
| 4 | D Davies | 4 | 51:24 |
| 5 | E Evans | 4 | 60:17 |
| 6 | F Frasier | 1 | 55:13 |
| 7 | R Richards | 5 | 131:01 |
| 8 | S Simons | 3 | 121:11 |
| 9 | T Taylor | 1 | 140:06 |
| 10 | V Vickers | 0 | Retired |

**Table 2.4: A typical set of results.**

CHAPTER 3

# PLOTTING BEARINGS

This chapter describes six ways of measuring and plotting the bearing of a hidden transmitter. They all have different advantages and disadvantages and the list is not exhaustive. The beginner should experiment with them all before deciding what works for him or her. It may be expedient to use one method on 3.5MHz and a different one on 144MHz or, indeed, different methods at different points in the same competition. The choice is yours.

## REASONS TO PLOT BEARINGS

There are three reasons why it is a good idea to plot some of the bearings of the hidden transmitters on the map given out at the start. See **Fig 3.1**.

- The competitor can then use the map with the bearing marked on it, to choose a good route to a hidden transmitter.
- Simply as an aid to remembering all the information being gained. For the M60s and W50s with data for just three transmitters to keep in mind, things are not too bad, but the M21s have to maintain spatial awareness for five transmitters and this is difficult for most people.
- Especially in competitions with a common start and finish, moving to the location of the first transmitter chosen, creates a baseline from which 'fixes' can be obtained on some of the others. If the bearings at the start and end of such a baseline are plotted, the approximate area in which one or more of the later transmitters needed, can be identified.

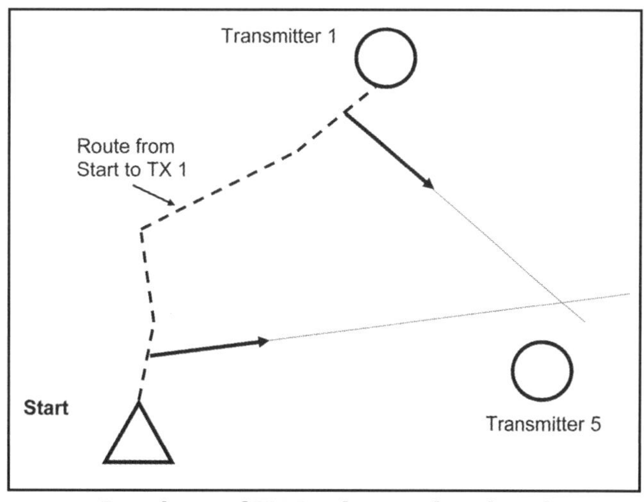


**Fig 3.1: Bearings of TX 5 taken on leaving the start and again while waiting close to TX 1 for it to come on the air, allow the approximate area of TX 5 to be identified.**

It is important to realise that the better you get at ARDF, the fewer bearings you need to plot. However, for the beginner, it is very important to plot bearings and to use them in conjunction with the map. As a basic piece of procedure, start your first competition by taking a set of bearings on all the transmitters you are hunting from the end of the start corridor.

**TAKING AND PLOTTING A BEARING USING A COMPASS WITH A RECTANGULAR BASE-PLATE**

Having decided the direction of the hidden transmitter by swinging the receiving antenna from side to side and listening to the strength of the signal, pick out a prominent object in that direction. Point the base plate of the compass in the direction of the object chosen and rotate the housing of the compass needle until the needle is aligned with the parallel lines on the floor of the housing (see **Figs 3.2(a), 3.2(b)** and **3.2(c)**). This has set the compass to that bearing and it is possible to read off the bearing on the centre line of the base-plate.

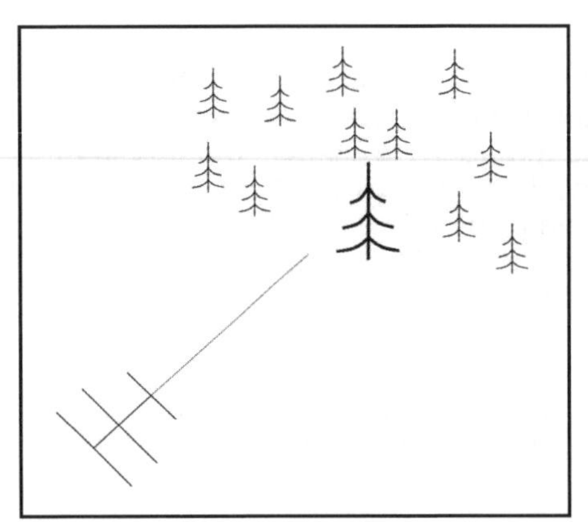

**Fig 3.2a**

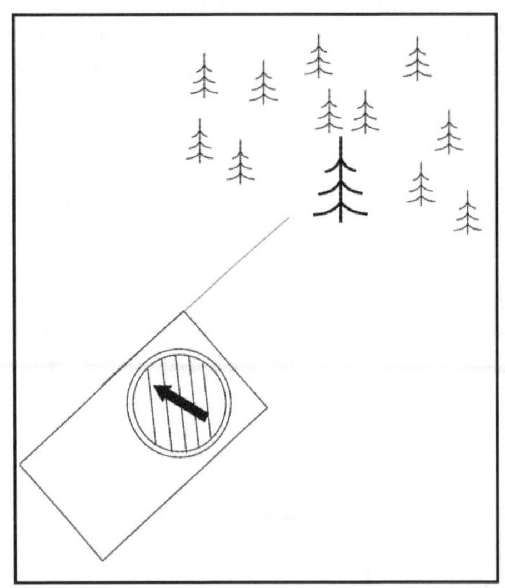

**Fig 3.2b**

**Fig 3.2c**

**Taking a bearing with a compass having a rectangular base-plate: (a) Swing the antenna to get signal maximum (minimum on 3.5MHz) and pick out a prominent object in line with the antenna; (b) Align the side of the rectangular base-plate with this object; (c) Keeping the compass aligned, rotate the compass needle housing so that the lines underneath the compass needle are aligned with the compass needle.**

There is no need to read off the bearing because the compass is now used as a protractor to transfer the bearing to the map. With the long edge of the compass base-plate through your current position of the map, rotate the whole of the compass until the lines on the floor of the needle housing align with the magnetic north arrows on the map. **Fig 3.3** refers. Now draw a line along the edge of the compass which passes through your position on the map. Note that it is normal practice for the maps used in ARDF to have lines of magnetic north drawn on them.

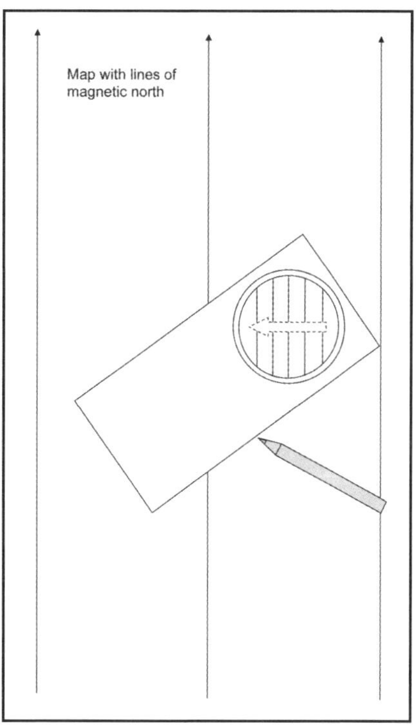


**Fig 3.3: Using a compass with a rectangular base-plate to plot the bearing on the map. The lines under the compass needle are aligned parallel to the lines of magnetic north on the map, with the side of the compass passing through the present position on the map. A bearing line is then drawn along the side of the compass. The position of the compass needle is of no interest during this part of the procedure, hence it is shown dotted.**

Now all this is a bit of a palaver. The competitor is juggling with a radio and its aerial, a compass, a pencil and a map. With only two hands, life gets difficult. To take and plot a bearing as described above, the competitor is usually forced to stop and put some of the things being carried on the ground. That said, it should be possible to take and plot a bearing in the 60 seconds that the transmitter is on the air (having practised the procedure beforehand).

**WITH THE COMPASS FIXED TO THE ANTENNA**

Provided the compass is positioned sufficiently far from any magnetic materials used in the construction of the antenna or the receiver, the compass can be fixed to the antenna and the compass bearing read straight off when it is pointed along the direction of maximum signal (for 144MHz) or signal null in the case of 3.5MHz. See the photographs. Experimentation is needed to determine a spacing at which the compass needle

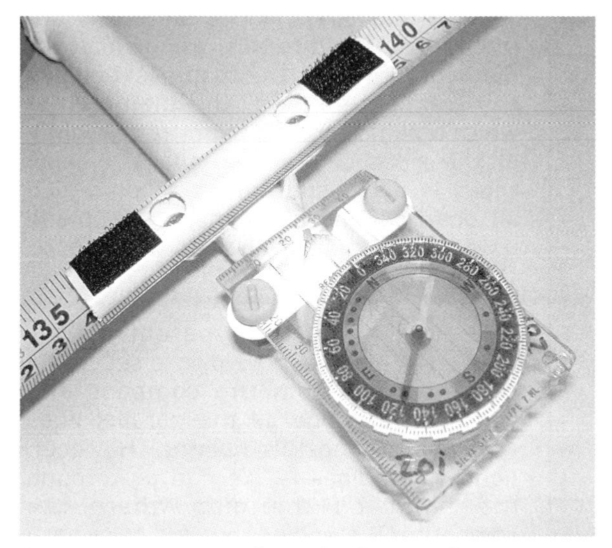

**A compass mounted on the boom of a 144 MHz antenna.**

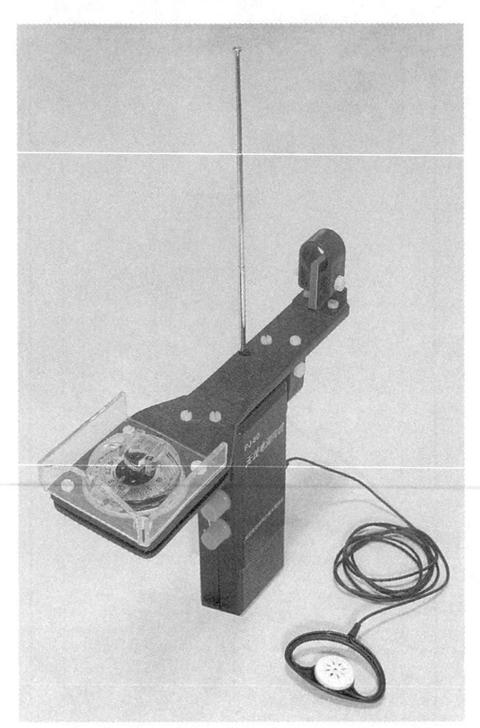

**A Plastimo sighting compass mounted on a PJ80 3.5MHz receiver. Plastic materials and nylon nuts and bolts were used in the vicinity of the compass.**

is unaffected by the presence of any ferrous materials.

There is still the problem of transferring the bearing to the map, but at least one source of error has been removed, and that is the error involved in picking out a prominent object in the direction of the transmitter and then taking a bearing to the object with the compass. Now, this part of the procedure becomes a single-step process of pointing the antenna and reading the bearing.

**Transferring the bearing to the map**

**Method 1**

To transfer the bearing to the map, a protractor can be used. In other words, there is a reduction in potential error, but no change in the number of things to be carried. The compass is now attached to the aerial, but a protractor is needed.

**Method 2**

This method is often used by competitors from eastern Europe. It uses a transparent disc marked with parallel lines. The disc is fixed to the map board by a rivet or bolt at the centre and is free to rotate.

Top layer – transparent sheet for plotting bearings

Transparent circular disc with parallel lines inscribed. Pivots about its centre.

Map fixed to map board

Map board with hole for bolt holding the circular disk

**Fig 3.4: Exploded view of the special map-board which facilitates the plotting of bearings obtained from a compass on the antenna.**

The map assembly now comprises four 'layers' (**Fig 3.4**). At the bottom is the map board. Fixed to this, using whatever method desired, is the map. Next comes the transparent disc, the diameter of which is sufficient for it to protrude slightly at the sides of the map-board. Finally, there is a transparent sheet on which all bearings and other marks are made using either spirit-based felt-tip or chinagraph pencil.

The edge of the transparent disc is marked in degrees. The disc is set to the bearing read from the compass and when this is done all the parallel lines now point along this bearing. However, none of the lines is likely to go through the point on the map where the competitor is standing, but it is quite easy to draw a freehand line parallel to the lines on the disc, which does go through the current position. This freehand line is drawn on the topmost transparent sheet and the map can be seen through the two transparent sheets (disc plus top sheet), as illustrated in **Fig 3.5**.

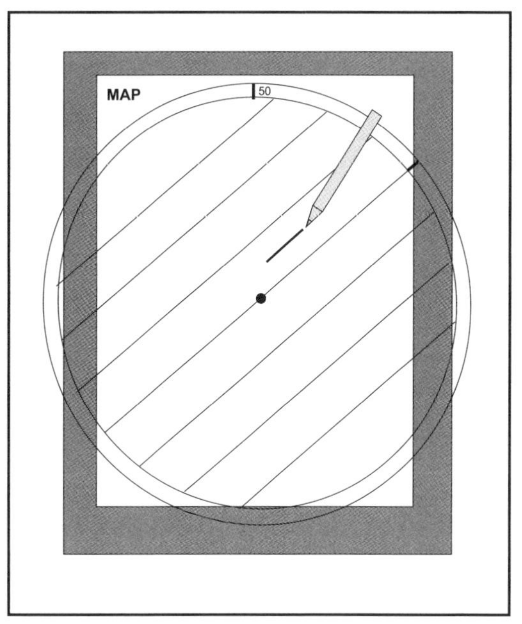


**Fig 3.5: Plotting a bearing of 50° read from the compass mounted on the receiver. The transparent circular disc is set to 50° at the top (note that the scale round the outside of the disc is anti-clockwise) and a freehand line is drawn through the current position on the map and parallel to the lines on the disc.**

## WITH THE COMPASS MOUNTED ON THE MAP BOARD

On 3.5MHz in particular, bearings are reasonably accurate and a set of quality bearings plotted at the start can be very valuable. If a compass is mounted on the map board, the board can be rotated so that the map is 'set' to magnetic north (ie the magnetic north lines on the map point to magnetic north in reality).

The receiver can be rotated for the sharp null experienced when it is pointing at the hidden transmitter and can then be brought into contact with the aligned map. The errors implicit in taking bearings by identifying a prominent object in line with the receiver antenna and then measuring and plotting the bearing to this object are all removed. The bearing is now plotted directly from the aligned radio on to the map. There are then two ways in which the bearing can be plotted:

### Method 1

Draw a line along the bottom edge of the receiver case (**Fig 3.6)** parallel to the ferrite rod (or at right angles to the loop). The disadvantage is that it is hard to juggle map board and radio and a pencil with only two hands. Consequently, it is all too easy to end up scrabbling about on the ground to get the line of the bearing drawn.

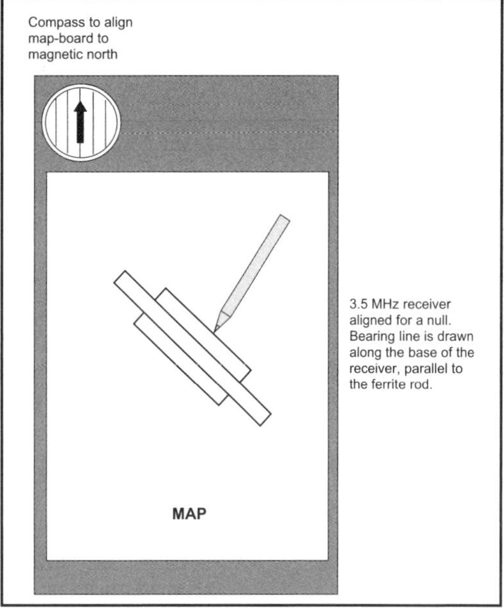


**Fig 3.6: Map board aligned to magnetic north and the side of the receiver used to draw the bearing.**

To overcome this problem, make an L-shaped bracket to bolt to the bottom of the receiver (see the photograph). The up-stand should be about 10 or 15mm and the flat part needs to be 30 – 40mm wide. The photograph shows such a bracket attached to a 3.5MHz receiver built into a Hammond case.

Holding the receiver in one hand and the map-board in the other, follow this procedure:

- Roughly align the map-board to magnetic north.

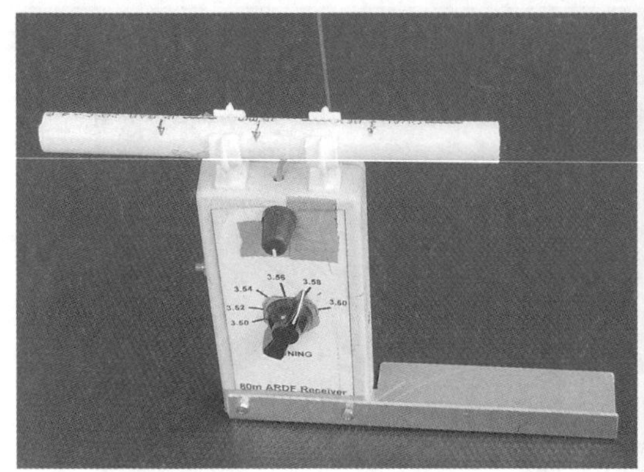

**The 3.5MHz receiver with an L-shaped bracket fixed to the bottom.**

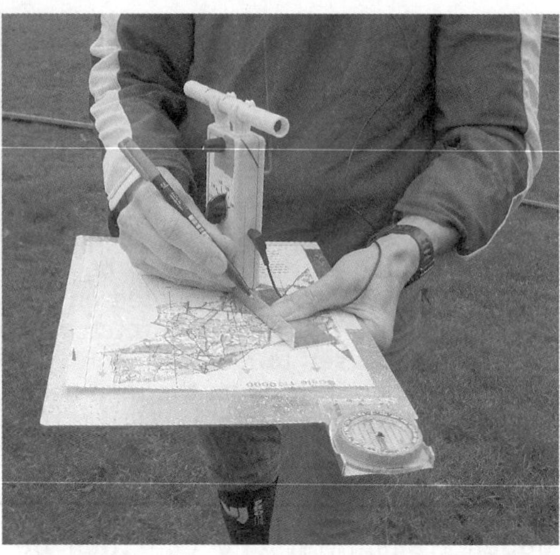

**The receiver is aligned to the null, the map-board is aligned to magnetic north, and they are clamped together using the fingers and thumb of the left hand, ready to have a bearing line drawn on the map. The flat part of the bracket facilitates this clamping action. The up-stand of the L-shaped bracket forms a straight edge along which the line can be drawn.**

**To plot a bearing line on the right-hand side of the map (which is out of reach if the receiver is held as in the previous photograph), the competitor now stands facing south and uses the left hand to clamp together the map-board and the receiver near to the eastern edge of the map.**

**A bearing line being plotted using a chinagraph (grease) pencil taped to the receiver.**

- Carefully rotate the receiver and find the null. Then hold the receiver firmly in this direction.
- With the other hand, carefully align the map-board with magnetic north and then 'offer it up' to the underside of the receiver keeping all the alignments. Make sure the edge of the bracket passes over the current location on the map. When in contact, grip the map-board *and* the flat part of the flange between the fingers

and thumb of one hand, as shown, while using the other hand to plot a bearing line on the map.

- The second photograph of this group shows this procedure for a point near the left side of the map. If the current location is over to the right of the map, keeping the map aligned in the same position, move round so that the map-board is to the south of your body. Then the left hand will be able to clamp the map-board to the receiver on the right hand side of the map. This is shown in the previous photograph.

**Method 2**

An alternative approach is to attach a downward-pointing chinagraph (in case it is wet) pencil to the receiver case. Felt-tip spirit pens are not very suitable since they dry out easily and there is every likelihood that one will cease to function during the competition if left with the tip uncovered, as required here.

With the map-board aligned to magnetic north and the receiver aligned for the null, place the tip of the pencil on the map at the spot where you are standing, and move the pencil along the line of the null. This is easiest with a ferrite rod receiver where the line of the rod gives the direction in which the receiver must be moved to draw the line of the bearing. This can be done without having to put anything down on the ground and with the minimal interruption to forward progress.

As with method 1, best results are usually obtained if the receiver is rotated for the null first and then the carefully-aligned map-board is offered up to the tip of the pencil so that they meet at the spot on the map where the competitor is standing, as illustrated in the photograph. Once contact is made, the receiver is moved carefully in the direction of the null.

The line plotted is usually a bit wobbly but it is a fast technique, which is especially valuable as the transmitter gets closer. Speed is then usually of the essence.

**WHY BOTHER WITH A COMPASS AT ALL?**

There are two important things to bear in mind at this point.

- Speed is of the essence (if you hope to do well) and stopping to plot bearings slows things up a lot.
- Especially on 144MHz, the direction of the signal maximum is never terribly accurate. Multi-path propagation and being caught in a poor location with a lot of screening when the transmitter comes on the air, are both factors which detract from the accuracy of the most carefully measured and plotted bearing. Why not accept that the 144MHz bearings are never going to be that good and proceed differently?

Consider this scenario. You are heading for transmitter 1 and are moving down a forest road in the general direction of transmitter 1. You know that transmitter 3 is away to the right. As you move down the road, transmitter 3 comes on the air. Swinging the 144MHz Yagi from side to

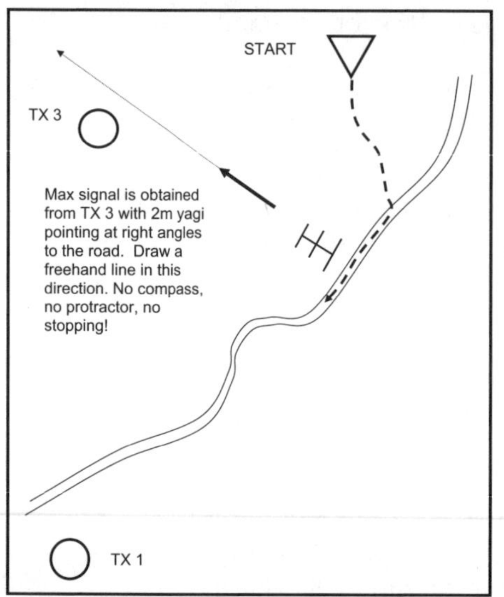


**Moving down the forest road, transmitter 3 transmits and the direction appears to be at right angles to the road. Draw a freehand line at right angles to the road and mark it with a '3' to denote transmitter 3.**

side, while moving along the road, gives a bearing as near as you can judge to be at right angles to the road you are on.

In this situation the 'bearing' of 90° to the right of the road can be marked on the map by eye alone without even stopping. Just draw a freehand line at right angles to the road. There is not a lot of point in measuring the bearing as, say, 88°, and then carefully plotting it with a protractor when the accuracy of bearings on 144MHz, even from a good bearing-taking location, is ±10° at best.

On 3.5MHz, the bearings will be significantly more accurate, but this procedure is still valuable by virtue of the time that can be saved. It is also possible to use linear features on the ground, such as a vegetation boundary or a path, as a guide to plotting bearings freehand. If the bearing seems to be 20° right of the line of a footpath being followed, then estimate the 20° and plot a freehand line.

A variation on the 'why bother with a compass?' theme is when the bearing taken passes through a clearly identifiable feature on the map. There is then absolutely no point in using the compass, just draw a line from the present position on the map through the feature identified on the map.

CHAPTER 4

# THE TOP 10 TIPS TO IMPROVE PERFORMANCE

This chapter sets out to give 10 hints by which performance can be improved. As the title implies, the list of tips is not exhaustive, but gives the 10 hints most likely to be a real help to the competitor in improving performance. The tips are arranged roughly in the order in which they might prove useful in a competition, starting with two to adopt before crossing the start line.

**Tip 1 – Always draw the exclusion circles round the start and finish**

The IARU rules state that there will be no transmitters placed within 750m of the start and within 400m of the finish. Also, the hidden transmitters must be a minimum of 400m apart. In local competitions using a small area of ground, it is not uncommon to find that the 750m rule for the start is replaced by 400m. This will be clearly stated in pre-race information.

In events where the start and finish are co-located, it is obvious that the greater radius of the start exclusion zone will over-ride the finish zone, so that there will be no transmitters within 750m of either the start or the finish.

The map will be given to the competitor only five or 10 minutes before the start in the major competitions. Assuming that the circles are not overprinted on the map, the only things marked will be the location of the start using a triangle and the location of the finish using a double circle. With such a short time constraint, the circles must be drawn in accurately and quickly.

Using a stencil is the way to do this and one can easily be made from a piece of thin card, maybe part of a breakfast cereal packet.

**Stencils (cut from an old cereal packet) for drawing the limits of the exclusion zones around the start and finish. The one on the left is for use with maps of 1:10,000 scale and the one on the right for 1:15,000.**

The first photograph shows suitable stencils. The outer edge is for drawing the start exclusion zone and must represent a radius of 750m for the map

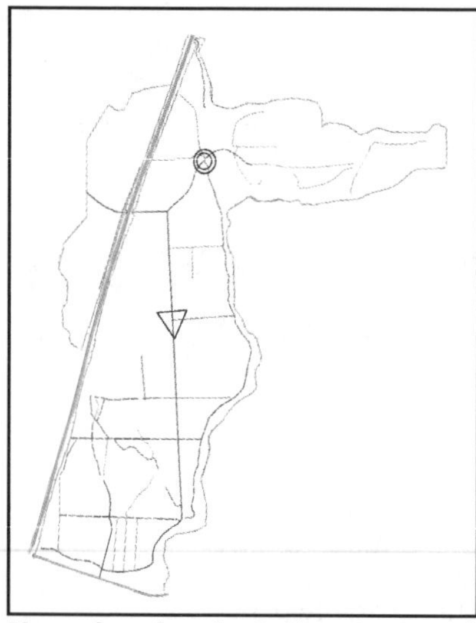

**Map showing just the start and finish marked in grey.**

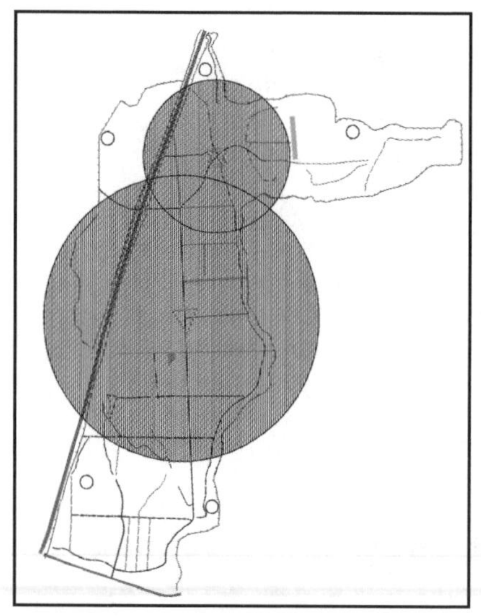

**Map with the exclusion circles marked and shaded.**

scale in use. Maps are normally either 1:10,000 or 1:15,000 with the former most popular for local events and the latter for big international competitions.

The inner circle corresponds to the 400m finish circle and the pointed part indicates the centre of the start triangle or of the finish double circle.

For a 1:10,000 scale, the radius of the outer circle is 75mm and the radius of the inner one is 40mm. For 1:15,000 maps, the stencil is smaller with an outer radius of 50mm and the inner one 27mm.

**Illustration**

To show how helpful plotting the exclusion circles can be, an example is taken from an actual race in England. Due to the restrictions imposed as a result of having a competition alongside an orienteering event, the start was near the centre of a long thin wood, with the finish nearer the northern end. The map issued to competitors had just the start and the finish marked. These are shown as the grey triangle and double circle respectively, on the first map.

As soon as the circles are added (second map), it can be seen that the 750m start circle cuts the wood into two halves. The finish is at the top but there is not enough room in the northern part to accommodate five transmitters each with their 400m exclusion zones. Hence some transmitters must be in the southern part of the wood.

As soon as one of the wanted transmitters is detected to the south of the start, the wise competitor simply 'legs it' down the main forest road to get beyond the 750m zone edge as quickly as possible. The wanted transmitters in the southern end of the wood are hunted down before the competitor heads for the northern part to find the remainder.

This is an extreme example and there would have been far better places to locate the ARDF start and finish had it not been for the orienteering event running at the same time.

**Tip 2 – How to write on wet maps**

Still in the five-minute window after being given the map and before starting, the map with exclusion zones marked on it must be prepared for the competition. If the weather is fine and dry with no indication of rain and the woods are also dry, then the easiest solution is to use the map unprotected. Most competitors will attach the map to some sort of light-weight board so that bearings can be marked on it.

However, in wet conditions, the map needs protecting and at the same time it must be possible to draw bearings on the map. Protection can be achieved by placing the map inside a clear plastic cover. The A4 plastic wallets designed for loose-leaf binders are a popular choice. However, the map must not be allowed to move around inside the plastic covering, otherwise bearings previously plotted can become inaccurate. The map may be a tight fit in the wallet but, if not, use bits of electrical tape to hold it in place.

An alternative method is to cover the map with TranspaSeal or Coverfilm. These are proprietary clear plastic film products that stick to the paper. Since it sticks to the whole of the map, there can be no movement between the two, but sticky plastic film is not the easiest thing to apply smoothly when in a hurry.

Writing on the plastic covering of the map can be done with either felt-tip pens or chinagraph (grease) pencils. There are two sorts of felt-tip pen, the water-soluble kind and the spirit pen. It is essential to use the latter if you opt for a felt tip. While the ink of a sprit-based felt tip will not run once it is dry, it cannot be applied to a wet surface without running. Hence the choice of most competitors is the chinagraph pencil, which will write on plastic in the wet and does not run. The leads of these pencils are very fragile and they cannot be sharpened to a fine point. Some competitors cut them down to half-length and attach them to a finger of the writing hand with a loop of elastic. One of the ever-present problems of ARDF is how do you carry all the stuff, and having the pencil attached to a finger is a partial solution.

**Tip 3 – On 144MHz head for a high spot**

A good set of initial bearings can be the key to a successful run, especially if you are a newcomer. On 144MHz, multi-path propagation can be a real hindrance when it comes to getting a decent set of initial bearings. If bearings are taken in a valley, they are rarely accurate and the signals often arrive along the line of the valley even if the transmitter is off to one side.

The solution is to look for a high spot with a clear take off in the direction of the hidden transmitters. Be on your guard for course planners who site the start corridor at the top of a steep slope. This was the case in the 2004 World Championships in the Czech Republic. The start corridor led to the steep sites of a very deep re-entrant. The temptation was to race off down the slope and end up in an area where good bearings were unobtainable. See **Fig 4.1** for an example.

**Tip 4 – How to recognise and deal with multi-path propagation on 144MHz**

On 144MHz, if there is a single large signal peak, the chances are that there is little multi-path propagation present and you may have a near line-of-sight path to the hidden transmitter. However, if there is more

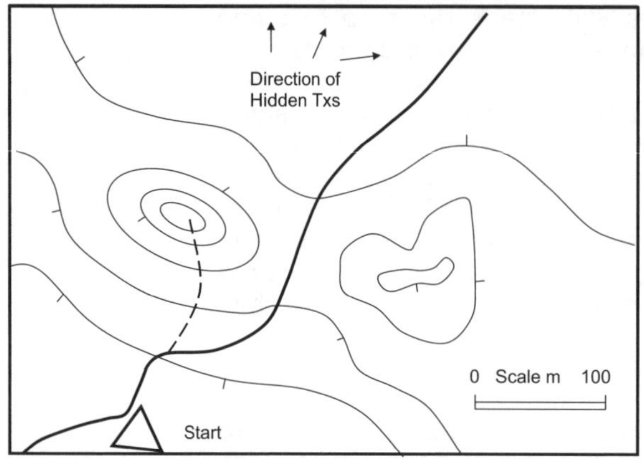


**Fig 4.1: The start is screened from the competition area by some high ground. This lies inside the 750m circle. The hilltop, at the end of the path that leads off the forest road, is an excellent location from which to obtain a set of initial bearings on the 144MHz band. (Note: the small check marks on the contour lines denote the lower side of the lines.)**

than one peak, and you are uncertain as to which is correct, the dreaded multi-path propagation is present.

Courses over essentially flat areas are unlikely to give very much multi-path, but as soon as the competition area is hilly then multi-path must be assumed as being probable at times.

An aerial, which has excellent side- and back-rejection of signals, is the first requirement. Most competitors use a three-element Yagi because this antenna can have a clean pattern with a boom length which is manageable in the woods. With such an antenna, it is easier to differentiate between the various signals.

Once out in the woods, if multi-path propagation is detected, *move* while the transmitter is on the air and see how the bearings you are getting vary, and how some of the signals arriving at your antenna rise and fall in strength. It is possible to form a view as to which of the signals is the main one and hence which bearing is to be believed.

It is also important to look at the terrain. If you are in a deep valley, it is probably a lost cause. The chances are that a multi-path signal is significantly stronger than the direct one and it is likely to arrive along the line of the valley.

The same effect can sometimes be observed in woods. Taking a bearing on a wide ride (an often-straight gap between forest blocks) results in a bearing directly down the ride when, in fact, the true bearing is off to one side. The ride appears to 'guide' the signal down its length.

Finally, remember that multi-path propagation is essentially a 144MHz problem where space-wave propagation occurs. Down on 3.5MHz, the propagation mechanism is surface-wave and very little multi-path propagation is observed here. When it does occur, the bearings are not disturbed very much.

**Tip 5 – When waiting for a transmitter to fire up, position yourself where you can move about easily**

There is no point being in the middle of some 'grot' when a transmitter comes on the air. Having set your watch carefully before you started the race, you will be able to predict when the transmitter you are hunting is next going to come on the air. Smart folk actually set their wrist watches so that they get to midnight at the appointed start time. This means that it reads your race time without you having to remember to press 'start' on a stop watch when you are hovering nervously on the start line.

Make sure you are on a track or ride. Better still be at a track crossing to give you four fast routes. When the transmitter fires up, run down the track nearest to the bearing you measure. Continue to take bearings as you run and see if the bearing starts to change noticeably. This is easier to judge if the track is a straight one, as **Fig 4.2** illustrates.

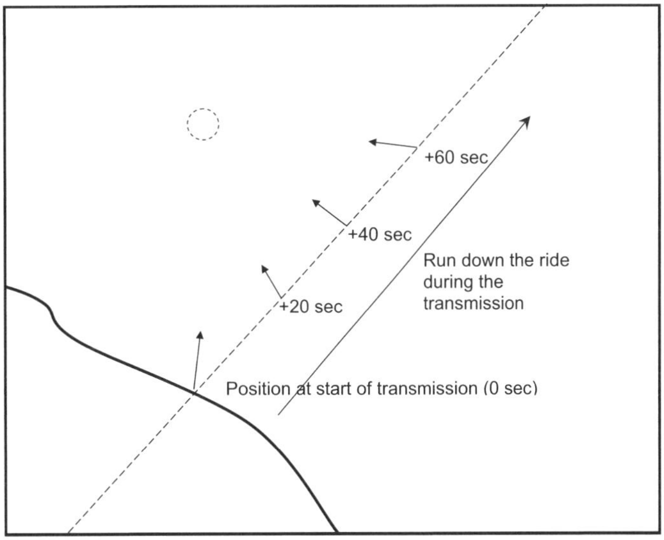

If the bearing does start to change, the rate of change enables an estimate to be made of how far the transmitter is to one side of the track. If the transmitter stays on the air for long enough, run the track until you are level with the transmitter and then dive in at right angles to the track for the distance you reckon the transmitter is away from the track.

**Fig 4.2: When the transmitter comes on the air, the competitor is at the forest road/ride junction. The competitor sets off in a north-easterly direction along the ride because this is the closest to the initial bearing (shown by an arrow). Further bearings to the transmitter are taken during the 60-second transmission. These bearings swing round more and more to the left . The competitor now has an idea of where the transmitter is at right angles to the track and how far into the trees it is located.**

### Tip 6 – Build up information about later transmitters while you hunt down the early ones

Beginners often think ARDF is conducted as it is depicted in many films, where two DF stations are usually positioned along a base-line roughly at right angles to the direction of the transmitter. The DF stations take bearings and the place where the bearings cross is the location of the transmitter. See **Fig 4.3**.

In practice, a competitor who moves from the start, at an angle to the bearing of the first transmitter in order to establish a base-line, will be beaten by someone who simply heads for the first transmitter and runs down the bearing. The latter individual can use range estimation techniques (see Tip 7) to help locate the transmitter.

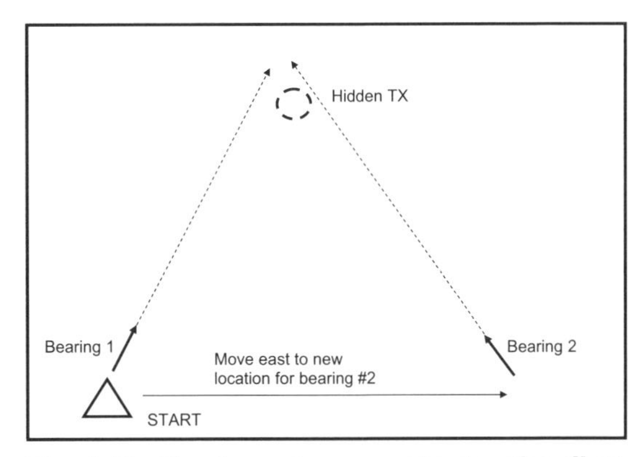


**Fig 4.3: Moving at an angle to the first bearing to establish a base-line. Once at the far end of this line, a second bearing will intersect with the first to give an indication of the location of the hidden transmitter.**

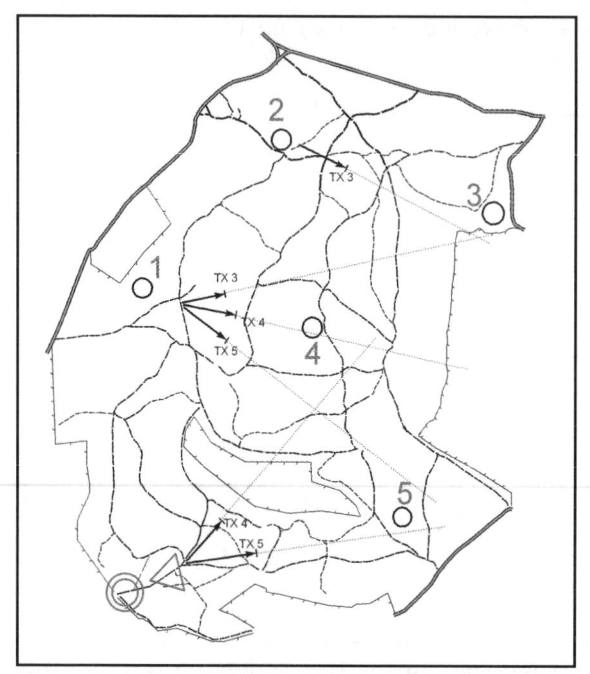


**Map of a hunt in which bearings taken *en-route* from the start to transmitters 1 and 2, enabled transmitters 3, 4 and 5 to be located within two or three hundred metres so that they could be hunted down easily and quickly on the way back to the finish. The competition used 144MHz, and the accuracy of the bearings is reasonable for this band.**

That said, in events which have a co-located start and finish, there is normally ample opportunity to use the route taken to the first and maybe the second transmitters as a base-line from which bearings can be taken of the remainder. This point was mentioned in Chapter 3 where the reasons to take and plot bearings were discussed. It is frequently the case as you get into the area of a hidden transmitter, that you have to wait for it to fire up before deciding the next move. While waiting, bearings can be taken on transmitters required later in the hunt and a picture built up of their locations.

**Illustration**

This example is taken from an actual DF hunt. From the start, it took one hour to locate transmitters 1 and 2. However, by the time the competitor arrived at transmitter 2, sufficient bearings of 3, 4 and 5 had been plotted to allow the remainder of the course (2 – 3 – 4 – 5 – Finish) to be completed in just 40 minutes.

**Tip 7 – Develop reasonable range estimation**

A trade-off has to be made between the accuracy of the range estimation and the complexity of the equipment needed and the time taken to use it. The IARU rules specify the transmitter power and, with fairly uniform antennas, range estimation is based on the received signal strength. The competitor needs to be able to determine with reasonable certainty whether the transmitter is 100m away, 200m, 400m, 800m and so on. In other words to an accuracy of a factor of two.

In the woods, the eyes need to be seeing where you are going, avoiding trees, holes in the ground and keeping to the path. If you hope to do well, there is no opportunity to keep looking at an S-meter. Hence, the ears must be the faculty used for range estimation.

This is achieved by crudely calibrating the receiver gain control in terms of distance to the transmitter. For what you, as an individual, define as a 'comfortable' volume of audio received from the transmitter, the receiver gain control should point to the correct range on a scale. To achieve this, access is needed to a transmitter with the usual aerial arrangement. The chapter on training and practice gives some advice on calibrating your receiver in this way.

On 144MHz, the accuracy is less good because transmitters can be screened by cliffs, hills and other topographical features. Additional information about such features is derived from the map and an assessment made of the likely accuracy of the range information.

The surface-wave propagation on 3.5MHz gives more predictable results, less dependent on the topography.

**Tip 8 – When the transmitter stops, keep running the bearing**
This tip applies when close to the transmitter, when the range estimation is down to 400m or less. A lot further away than this, it is not usually the best tactic to run directly at the transmitter but to look for paths and tracks leading more or less in the right direction so that you can get into the general area of the transmitter as quickly as possible.

If you think you are getting close and the transmitter goes off, transfer the last direction you have got to your compass and then 'run the bearing'. On 3.5MHz especially, the sharp null of the ferrite rod gives a very good direction to follow.

In big international competitions, where split times are available, it is an education to see just how many competitors find a transmitter after it has stopped sending. Two or even three minutes is not unusual and it is not always the result of searching but more often it is the result of 'running the bearing'.

**Tip 9 – Attenuate and tune**
The characteristics of receivers differ, and one of the keys to success is to be comfortable and experienced with the receiver you are using. As you hunt down a transmitter on an AM receiver with no AGC, the volume rises as you get closer and closer to the transmitter.

It is a characteristic of human hearing that we are less able to distinguish different signal levels when their volume is high. To maintain the ability to tell the direction in which the signal is strongest (or weakest if using the null on a 3.5MHz receiver) then turn down the volume as you get close.

Depending on the receiver, when the received signal is loud, the receiver may overload resulting in a reduction in volume. When the aerial is swung to one side, the received signal drops, the receiver ceases to be overloaded and the volume *rises*! Hence the hapless competitor is convinced that the direction of the hidden transmitter is 20° or 30° away from the correct bearing. The solution is just the same – keep turning back the gain to maintain a comfortable volume. In addition, on 144MHz, where all transmitters may not be exactly on the same frequency, make sure the receiver is correctly tuned to the transmitter you are hunting.

**Tip 10 – Always check the sense on 3.5MHz and all round on 144MHz**

Consider this scenario for a moment.

It is a warm sunny day. You have a good bearing for the next 3.5MHz transmitter that you need. The forest road in front is sloping gently down hill and it leads straight along the bearing you have measured to the transmitter; you lengthen the stride and 'go for it'. You are on a roll and each time the transmitter comes up, you check the direction and it is always straight on down the road... that is, until you notice it is getting a bit weaker and check the sense. The transmitter is now behind you. It is located close to the forest road and you ran past it while it was not transmitting and *you forgot to keep checking the sense*!

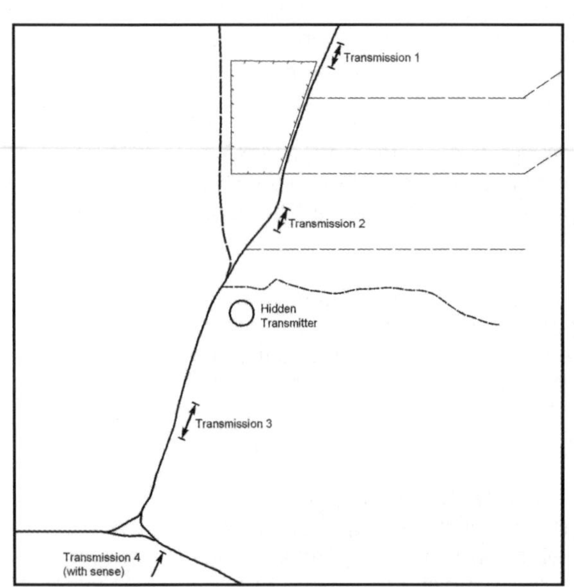


**The arrows show the bearings taken when the transmitter was sending. Due to the 180° ambiguity on 3.5MHz, the competitor ran straight past and continued down the road, believing the transmitter was still in front.**

Bearing 3 taken by the competitor using a ferrite rod or loop antenna, was to the rear and not further down the road in front. Having got to the location where transmission 4 was received and the sense was used, there was nothing else for it but to trail back up that gently sloping forest road, regretting that the sense had not been checked every time the transmitter came on the air.

The moral of the tale is always, *but always*, check the sense on 3.5MHz – assume nothing.

On 144MHz, there is no 180° ambiguity to lead you into the kind of error seen in the above example but multi-path propagation obliges every experienced competitor to swing the aerial right round at least once during each transmission just to check that the signal is not stronger 'off the back'.

CHAPTER 5

# ORGANISING AN EVENT

Events are the lifeblood of ARDF. Without a programme of events each year, the sport would just wither and die. Those who are prepared to organise events ensure the continuation and development of the activity.

Organising an event is not as difficult as some might believe and KISS (keep it simple, stupid) is a good maxim. If the process is allowed to be too complex and time-consuming, organisers would soon find it difficult to give the time necessary. By focusing on the essentials, local events can be organised by just a couple of people and, in some cases, by just one individual. This chapter sets out to give some guidelines and ideas for event organisation.

There are many similarities between ARDF and orienteering. More information on orienteering event organisation and example forms and checklists can be found at
**www.harlequins.org.uk/documents/HOC_Event_Manual.pdf**, and
**www.harlequins.org.uk/documents/HOC_Event_Manual_Appendices.pdf**

If you are organising a major event, further guidance can be had from the IARU Region 1 website 'Rules for Championships' page, **www.ardf-r1.org/html/ardf_rules.html**, which covers the technical and other issues associated with ARDF.

**Stand-alone or with an orienteering event?**
Advantages in teaming up with an existing or proposed orienteering event include not having to arrange land permission, parking, toilets and SI/Emit equipment. You may even be able to use the same start, finish and timing systems – greatly reducing ARDF manpower requirements.

Disadvantages include the need for careful liaison, limited choice of dates and venues and restrictions on the location of transmitters (not close to orienteering controls).

It is sensible to split the tasks of 'organiser' and 'planner' as follows.

The *organiser* is in overall charge and looks after everything except the courses and the siting of the transmitters. The organiser has ultimate responsibility for the event – including safety.

The *planner* is responsible for setting up the transmitters (including spares), checking that the batteries are charged and that the transmitters are programmed and synchronised. The planner chooses and checks the location for each transmitter. The planner sets up and takes down the transmitter equipment (transmitter, antenna, banner, punch or EMIT/SI unit) and checks that the unit or punch is functioning correctly. During the competition, the planner monitors the transmissions, ready to rush out with a synchronised replacement in case of failure.

A perfectly satisfactory local event can be staged by just two people. The organiser runs registration, and the co-located start and finish, while the planner sets out the hidden transmitters and reacts to any unplanned equipment failures. There have even been instances of a single individual running a one-frequency band event on a small area. If the organisation can be kept simple it is less of a chore to stage an event.

**Land permission**

In the UK it is almost always necessary to obtain specific permission before hosting an ARDF event. This can be complicated and can take a lot of time – there may be several landowners and other interested parties to be contacted. Certain areas are very heavily used for organised events – shooting, mountain biking, running, dog trials and so on, and it will be necessary to negotiate dates and positively inform other users. On at least one occasion an event has clashed with an informal 'free-range archery' competition – the event had to be abandoned. There may also be a land access fee – there are national orienteering agreements with the forestry commission and other bodies – these may be more favourable than what can be individually-negotiated. If the land is a nature reserve or incorporates a Site of Special Scientific Interest, permission will need to be sought from Natural England, Countryside Council for Wales, or Scottish Natural Heritage.

Unless you already have an existing contact with the landowner(s) it is best to work with the local orienteering club. It is likely that the orienteering club will know who to contact, will already have access agreements in place, and will be aware of orienteering and other activities planned for the area. Often an approach through the orienteering club will be better received and is less likely to confuse a busy land agent.

The landowner may require evidence of insurance – this can be obtained from the RSGB. In any case you should familiarise yourself with the details of this insurance so that you can be sure your planned activities are covered. The RSGB insurance for Affiliated Societies provides £5m of public liability cover.

If you are piggybacking your event on to an orienteering race – make sure that the landowners have been made aware that there is a radio element in addition to standard foot orienteering and exactly what this means.

**The map and area information**

You will normally obtain copies of the map from the local orienteering club. In certain cases, typically country and forest parks, the landowner may hold map stocks. The orienteering club may supply you with a computer map file – allowing you to print your own copies – but make sure that you obtain advice on colour settings and ink and paper selection.

The orienteering club may have an area information pack which will cover:

- access arrangements and key holders;
- a map showing area available for competition marked up with out-of-bounds zones;
- known hazards;
- areas with special environmental or other restrictions such as dog bans; parking locations;
- contact details for landowners, tenants and other users, including those who should be notified before the event takes place;
- the nearest police station and Accident & Emergency unit.

If no pack is available, make sure that you have asked sufficient questions of the landowner to ensure you have covered the necessary points.

**Safety and risk assessment**

Some landowners may require a risk assessment form to be provided to them as a condition of access. In any case it is good idea to carry out a simple assessment, with input from the planner who should be familiar with any anticipated hazards.

**First aid**

It is not normally practical to provide formal first aid cover for ARDF. It might be prudent to display a notice in the registration area notifying people of the absence of a first-aider. If possible identify someone with current first aid training, who will provide a simple first aid kit. Plenty of water and a survival bag or thermal blankets should be on hand in case of hyper- or hypo-thermia.

The organiser should know the location of the nearest available casualty hospital, nearest working telephone (if there is no mobile phone coverage), and how to gain vehicular access to remote parts of the land, with a key to open any locked gates.

**Dangerous features**

These should be taped off with yellow tape: this is the responsibility of the planner in the competition area and the organiser elsewhere. Where roads are crossed, warning signs for motorists should be put out. Traffic marshals may be required.

**Clothing**

Extreme weather conditions and/or exposed terrain may require cagoules to be worn or carried, and advance notice of this possibility should be given.

**Whistles**

The organiser should decide whether or not to enforce 'No whistle, no go' for competitors. If so, this should be clearly indicated to competitors on the information sheet, in the assembly area, and at the point where competitors set out for the start.

**Missing competitors**

Although people take part at their own risk, efforts must be made to ensure no one is left in the forest. At events with manual punching, it is usual to collect stubs at the start and match them with the control cards of people finishing. At events using electronic punching, a list of competitors who have not reported to the finish can be produced by the finish team.

Neither system is foolproof – both depend on everyone who starts reporting to the finish. The requirement to go through the finish is printed on control cards (if used) but also should be emphasised by notices. An additional safety check is to use the 'buddy system', with people being reported missing by their travelling companions and people on their own invited to leave car keys at Registration to ensure they report back. (Beware, however, of liability issues should the keys 'go missing')

If it seems likely that someone is left in the forest, enquiries must be made to get as much information about the person as possible with a view to mounting a search. Be prepared for this (with torches in winter) and ensure that sufficient help is available.

**Incidents**

If there any incidents, injuries, accidents or damage caused which might result in a legal claim being made, the organiser should make sure a contemporaneous record and kept safely for future reference.

**Liaison with orienteering organiser, planner and controller**

If your event is to be co-hosted with an orienteering event, as soon as the orienteering organiser and planner have been appointed, make absolutely sure that they are aware of the ARDF event and that they understand what ARDF does and does not involve. It is good idea to brief the controller as well.

The organiser is responsible for event safety, the planner for the courses and the controller for checking the courses for accuracy, fairness, safety and other, technical issues.

From the organiser, find out the location of the car park, registration, start and finish. The planner can tell you about restrictions and hazards and where exactly all the orienteering controls are to be positioned. It would be a very bad idea to have a transmitter located close to an orienteering control, and especially on a similar looking feature.

The orienteering information sheet should mention that ARDF will be taking place.

It is a very good idea to re-confirm the location of orienteering controls and ARDF transmitters on the day of the event – there may have been last-minute changes.

**Pre-event information and on-the-day information sheets**

The following information should be provided, either in the pre-event advertising or on a sheet to be handed out on the day or both – as appropriate.

Name of organising body.

- Date.
- Area name, location of the nearest town and map reference of car park.
- Travel directions (including where signposted from).
- Type of event and bands (ARDF or FoxOring, 2m, 80m or both).
- Any deviations from or suspension of standard IARU ARDF rules.
- Transmitters to be sought by class.
- Time limit.
- Map details, eg scale, special details, date.
- Type of terrain.
- Whether electronic punching will be used and, if so, arrangements for hiring chips.
- Whether pre-entry is necessary, arrangements for pre-entry including closing date.
- Whether there is entry on the day.

- Whether the event is suitable for beginners.
- Registration open from-to /start times from-to.
- Course closing time.
- Entry fees – juniors, seniors, parking charge.
- Facilities – toilets, first aid, refreshments, etc.
- Whether dogs are allowed.
- Safety statement 'All competitors take part at their own risk and are responsible for their own safety'. A note of any particular hazards and a safety bearing.
- On exposed areas or in winter whether whistles and cagoules are compulsory.
- If there are dangerous or environmentally-sensitive areas to avoid and whether these will be taped. A note on out-of-bounds areas that are not marked on the map: 'All fields out of bounds, roads are not to be crossed or run along'.
- Data Privacy Statement: The personal data you give will be used by the event organiser only for the purpose of processing and publishing entries and results. Entry data will not be retained after three months from results publication or the conclusion of any protest or appeal.
- URL for results.
- Officials – planner.
- Organiser – name, address, telephone number and e-mail address.

A sample on-the-day information sheet appears in Chapter 1, Fig 1.2.

**Registration sheets**

Each competitor should fill in a registration sheet with the following information.

- Name
- Call Sign or British Orienteering Federation Number
- Club or Society
- Class Entered (if 'running up', this should be declared)
- Car registration number

and then sign that they understand: 'All competitors take part at their own risk and are responsible for their own safety'. A sample registration form is shown in **Fig 5.1**.

After registering, the competitor is allocated a start time.

Declaration

1. I wish to enter the ARDF competition today.

2. By signing this form I agree that I take part entirely at my own risk. I agree that I shall not hold the officers and members of the Radio Society of Great Britain, the organisers of the event or the landowner responsible for any loss, damage or injury incurred by me during the event.
I accept responsibility for my own safety.

**Name in caps:** ........................

**Course/class:** ........................

**Car registration:** ........................

**Name of your Club:** ........................

**E-Mail address:** ........................

**Signed:** ........................

**Date:** ................

**Fig 5.1: Example of a registration form.**

**Organiser's/(planner's) timetable – for a stand-alone event**

| Time | Action |
|---|---|
| ASAP | Ensure that land permission has been obtained and that the area information is complete.<br>Ensure that the event is listed in the relevant fixtures lists. |
| 3 months before | Planner – obtain a copy of the map and visit the area, checking for suitable transmitter sites and any unexpected changes to the terrain (tree felling, new fences, seasonal vegetation changes).<br>Liaise with the planner over car park, registration, start and finish locations.<br>Decide whether event is using electronic or paper-punching.<br>Prepare publicity and arrange for publication on relevant websites and circulation to local radio societies.<br>Contact local orienteering club and agree arrangements for supply of maps and any loan equipment.<br>Prepare a task and equipment check-list. Confirm that you have sufficient manpower arranged for the tasks and that the supply of equipment has been arranged. |
| In the week before | Confirm you have all the maps and equipment. Ensure that transmitters are fully charged and working and that the planner is familiar with their operation.<br>Confirm that your helpers are still available and that they understand their duties.<br>Prepare information sheet, registration forms and map corrections.<br>If necessary, brief local residents and other area users.<br>Planner – a 'just in case' visit may be in order. |
| The day before | It can take a very long time to set up a transmitter – it is often best to position the support strings for the aerials in the trees the afternoon before. This also avoids a last-minute panic if it becomes clear that the intended sites are no longer suitable. |
| On the day | **Before the first start**<br>Planner – synchronise start, finish, EMIT/SI and transmitters – normally to GMT or BST as appropriate for the season.<br>Planner – revisit transmitter sites, putting out SI/Emit units, transmitters and aerials. Checks the punch pattern/correct function of the unit.<br>Put out direction signs, tapes, tents and other equipment.<br>Brief helpers.<br>**During the competition**<br>Planner – monitor transmitters, standing by to rush out with a spare in case of failure.<br>Monitor smooth running of the event.<br>**After the competition**<br>Planner – collect in transmitters, banners, etc.<br>Confirm that everyone is out of the forest – including those 'collecting in'.<br>Clear up and pack away |
| The week after | Write letters of thanks to landowners, etc.<br>Pay any fees due.<br>Return keys, equipment and any unused maps.<br>Update Area Information Pack |

**Sample Timetable**
If there have been incidents, inform RSGB / orienteering club and make sure records are retained – just in case!

**Equipment check list**

- Organiser's equipment check-list (most may be borrowed from your local orienteering club).
- Radio controlled clocks – 2.
- EMIT/SI units, hire chips and computers.
- Control cards.
- Cash float and box.
- Registration slips (and box).
- Information handouts.
- Start lists.
- Finish record.
- Shelter for registration (could be a vehicle).
- Tapes for start and finish boxes and funnels.
- Signs – Road, Registration, Start, Finish.
- Mobile phone numbers for planner, organiser and helpers.
- Maps pre-marked with start and finish.
- Spare whistles.
- Results display.
- 6 Transmitter sets* – 2m.
- 6 Transmitter sets* - 80m.
- String, catapults, tape, scissors, hammer – as required for setting up transmitter site.
- First aid kit and blankets / survival bag.
- Area Information pack and a copy of letter of permission – just in case. Tenants and rangers may need evidence and landowners may forget.
- Water.
- Pens and paper.
- Tables and chairs.
- Lunch!

* Transmitter (programmed and charged), antenna assembly, banner, stake for banner, ident for banner ('FOX-1 2m'), EMIT/SI unit or punch, waterproof bag.

**Planning: course setting for ARDF**
IARU-style ARDF is different from what you may have experienced in local T- or fox-hunts. The IARU rules state:
'2.2.2 Fairness
Fairness is a basic requirement in competitive sport. Unless the greatest care is taken at each step of course planning and transmitter placing, luck can easily become significant in ARDF competitions. The siting referee shall consider all such factors to ensure that the contest is fair and that all competitors face the same conditions during every part of the course.

'2.2.3 Competitors' enjoyment
The popularity of ARDF can only be enhanced if competitors are satisfied with the courses they are given. Careful course planning is therefore necessary to ensure that courses are appropriate in terms of length, physical and technical difficulty, transmitters siting, etc.
In this respect, it is particularly important that each course is suitable for the competitors doing that course.'

People may have travelled a long way to your event and it is important that they find the courses fun, satisfying and fair.

Aim to challenge rather than to trick competitors. You should encourage competitors to develop their ARDF skills through thought, experiment and practice and the course design and transmitter placement should reward competitors who effectively put these skills into action.

Random chance should not decide where a competitor comes in the results table. Luck will always have an influence - but it should not dominate. We should reward those who 'make their own luck'. Small errors (in bearings taken, route choice, route execution) should not result in big losses in time. In the absence of other information, a lucky random guesser should not be rewarded with big advantage over someone who guesses differently.

None of this means you cannot set the competitor a challenge or a puzzle but it must be fair.

If you saw the transmitter from 100m away only because someone was standing by it - *not fair*.

If you correctly DF-ed to within 50m of the transmitter and then spent 20 minutes searching - *not fair*.

**The course and transmitter locations**
You will need to select locations for a start, five transmitters, the beacon and the finish. There are no hard and fast rules - here are some guidelines.

- It is acceptable, and convenient, to have the start and finish in the same place - but this does reduce some of the thinking element of the competition.
- The IARU standard states that no transmitter will be situated closer than 750m from the start and no closer than 400m from any other transmitter (including the beacon). In small areas, it is normal to reduce one or both of these figures.

- The standard time limit is 120 minutes – in small areas and where two races are conducted in the same day this, and the number of transmitters to be found, may be reduced.
- Different classes seek different transmitters – suggested 'ideal' course lengths for informal races would be 6km for M21, 5km for M20/M40/W21, 4km for M50/W35/W20 and 3km for M60/W50. (10m of climb counts as an extra 100m 'level' distance.) It will, however, be difficult to achieve such an even division in practice, and often the higher number of transmitters to find for '21' makes sufficient difference even if overall length is similar to '50'.
- Note that requiring competitors to fight through 500m of dense undergrowth or crawl up and then tumble down steep hillsides will not make the planner popular.
- Placing a transmitter close to an impenetrable boundary is not a good idea if competitors are forced to 'choose a side' and commit before sufficiently discriminating bearings can be obtained. This is particularly true for fences, railways, motorways, rivers and deep cuttings where the amount of metal or water around and the shape of the terrain make bearings unreliable and crossing points are infrequent.
- When locating transmitters, consider the order in which they will be first heard. It is not a good idea to have 1, 2, 3, 4 all in the same direction and then 5 'behind' the start and away from the finish.
- Competitors should be able to have a good stab at selecting a reasonable (if not optimal) sequence within a couple of transmission cycles. They should not be forced to take a blind guess and commit to a strategy before information is available. A competitor who quickly grasps what is going on should be rewarded but someone who runs off toward a loud but distant transmitter 1 (hilltop?) only to find that they have already passed a close but quiet transmitter 3 (deep valley?) should not be unduly punished.

**The exact transmitter site**

The IARU rules say

'25.10 The flag shall be close to the transmitter antenna but not further away than 4m. The flag shall be visible to competitors when they reach the transmitter antenna'. And 'Flags should be positioned such that competitors may see them when they have reached an area about 10m around the transmitter. For fairness, the visibility of the flag should be the same whether or not there is a competitor near it'. And 'There shall be no objects disturbing the electromagnetic field and therefore obstructing correct measuring in the vicinity of the transmitters.'

Having DFed to the area in which the hidden transmitter is located – determined by means of cross-bearings or 'change in signal strength' or a combination of these techniques – the competitor should reasonably expect to find the transmitter after 'a bit of running around'.

This is not to say that transmitters should be visible from a long way away – just that it should be capable of being found by someone who correctly 'knows' that it 'must be near' within a couple of minutes of getting close to the transmitter site. In international competition, most transmitters are located within the minute following the end of a transmission.

Here are some examples of OK and not-OK sites.

| | |
|---|---|
| Not OK | In a small pit in open forest and the pit is not on the map. Finding the transmitter is a matter of chance – you don't know to hunt for it in a hole in the ground. |
| OK | In a large depression in open forest and marked on the map – where a good DFer might think, "I'll go and have a look in that". |
| Not OK | In a marked pit, but it is one of 30 in the same area. |
| OK | In a small area of denser forest "I know it must be somewhere in there!" |
| Not OK | 'inside' a small holly bush or thicket, where you could go right past and not see the banner. |
| OK | Provided it meets the other criteria, 'behind' something marked on the map – a wall, a crag foot, a large bush, a change in vegetation from 'low visibility' to 'good visibility' – as long as the transmitter is visible once you have passed the feature *and* where most competitors will approach from the 'hidden' direction. |
| Not OK | So close to something small and not marked on the map (a tree trunk, a clump of bracken) that the flag is hidden from competitors approaching from a wide sector of direction. As the feature is not on the map, the DFer does not know to 'look behind it'. |
| Not OK | In an area of dense summer or other ground vegetation – finding the flag is subject to random chance. |
| OK | In a marked ditch or gully – as long as the competitor can clearly see a good way along the ditch! |
| Not OK | Someone at the transmitter makes it visible from 100m away. |
| Not OK | In vegetation where 'elephant tracks' will quickly develop and the flag cannot be seen from 'outside' the area of vegetation. |
| Not OK | A single way into and out of a transmitter, departing competitors leading in other runners. |
| Not OK | Aggressive vegetation – brambles, nettles and the like. |
| Not OK | Where a very tall or very short competitor has an advantage – a control in a dense holly thicket where it is visible to someone small - or in bracken where a six footer can easily see the flag. |

Generally speaking, transmitters should be sited in areas where competitors are able to manoeuvre freely while the transmitter is on the air. This makes it easier for them to observe changes in bearing and signal strength while moving around.

Once you have selected and visited the transmitter site, it is a good idea to mark the exact intended position of the banner clearly with a length of tape. This means the site is easy to find on the morning of the event and you will not have to spend time worrying about where to put the banner. Remove the tape before the event starts!

In large areas or those with significant relief you should carry out a signal propagation test – ideally all transmitters should be audible from the starting point. If this is not possible, it should be clear from the location of the start, finish, and competition boundary, in which direction the competitors should run in order to pick up the transmissions.

CHAPTER 6

# TRAINING AND PRACTICE

The aim of this chapter is to suggest some activities *away* from competition that will improve performance *in* competitions.

## PHYSICAL FITNESS

People come into ARDF from two main backgrounds – the amateur radio world and orienteering. Those who come with an amateur radio background are often unsure of their physical abilities and usually over-rate the physical challenge of a competition. ARDF is not as strenuous as orienteering since competitors, especially inexperienced ones, tend to spend a lot of time standing around waiting for the wanted transmitter to 'fire up'.

That said, those who become enthused with ARDF usually wish to improve their physical abilities in parallel with their direction-finding ones.

### Health screening

In order to ensure that it is safe for you to begin a formal exercise programme, you should complete a simple health-screening questionnaire. Some simple questions are given here.

Has a doctor ever said that you have a heart condition?
Has a doctor ever said that you have high blood cholesterol?
Has a doctor ever said that you have high blood pressure?
Do you have diabetes?
Do you ever have chest pains?
Do you suffer from dizziness or fainting?
Have you ever had a bone, joint or back problem?
Do you have any other medical condition such as osteoporosis, epilepsy or respiratory disease?
Are you pregnant or have you had a baby in the last six months?
Is there any other physical factor that may limit your ability to exercise?

If you answered 'yes' to any of the questions above, you should consult your doctor before commencing a new exercise programme. In addition, it is advisable that all men over 40 years of age and women over 50 years of age consult their doctors before starting a vigorous exercise programme. A vigorous exercise programme involves exercise that makes you sweat and become slightly out of breath.

**Recommended amount of physical activity for health benefits**

As a minimum, the UK Health Education Authority recommends that everyone should carry out

- At least 30 minutes of moderate-intensity activity on five or more days of the week. It is not necessary to do the 30 minutes all in one go, it can be split up throughout the day.
- In addition, it is also recommended that, twice a week, individuals undertake physical activity that promotes strength and flexibility.

*Moderate intensity activities* include vacuuming, window cleaning, lawn mowing, raking, spring cleaning, golf, cycling (easy) and brisk walking. These activities make you breathe harder than normal and become slightly warmer.

*Vigorous intensity activities* include carrying heavy loads, shovelling, stair climbing, running, football, basketball, hill walking, aerobics and fast cycling. These are activities that make you sweat and become slightly out of breath.

*Strength activities* involve working the muscle against a form of resistance and examples include digging, lifting, carrying shopping, climbing stairs, body weight/ resistance activities, and exercises using a multi-gym.

**Flexibility activities** include gentle reaching, bending and stretching of muscle groups. Examples include gardening, housework, pilates, yoga and maintenance / developmental stretches.

**To gain fitness benefits**

If you would like to improve your fitness, you will need to carry out a more formal exercise programme. The physical activity that you perform will need to be at a higher intensity than that required for health benefits, and will also need to be progressed as your body begins to respond to the demands being placed upon it.

Activities should also be tailored to the component of fitness that you wish to improve. Here are some examples.

- Jogging / cycling (aerobic endurance - the number one requirement for ARDF).
- Resistance machines (strength / muscular endurance).
- Pilates / yoga (flexibility).

**Starting point**
For individuals who currently do no form of physical activity, the starting point of a programme would be at least 30 minutes of moderate intensity activity on five or more days of the week. Go back to the definition of moderate intensity activity above to check out how much of your existing activities qualify. If you are already moderately active and want to improve your fitness, a more formal exercise programme should be planned.

Lifestyle changes that can help to improve your health and fitness
Listed below are examples of lifestyle changes that individuals can make to help increase the amount of daily physical activity that they carry out.

- Always take the stairs, not the lift.
- Park your car a little further away from work and walk the remaining distance.
- Get off the bus one stop early and walk the remaining distance.
- Cycle to work.
- Spend 15 minutes in the garden on return from work; the fresh air will do you good and the garden will benefit as well!
- Walk to the local shop to pick up a paper; don't drive.
- Walk the dog twice a day.
- Go out of the office for a 10-minute walk at lunchtime. You will feel more refreshed on your return.
- When the children go swimming, don't sit and watch them, join in.
- Spend 10 minutes in the morning doing home exercises before going to work.
- Spend 10 minutes in the evening doing some stretches in front of the television.

**Exercising at home**

Exercising at home is very convenient. **Table 6.1** illustrates example activities that can be done at home to improve the components of health-related fitness.

| COMPONENT OF HEALTH-RELATED FITNESS | ACTIVITY THAT CAN BE DONE FROM HOME |
|---|---|
| Flexibility | Stretches |
| Aerobic endurance | Cycling outdoors<br>Indoor cycling<br>Jogging<br>Brisk walking |

**Table 6.1: Home exercises.**

**Stretches**

Stretches should be done as part of the warm-up and cool-down. Stretches can also be done to improve flexibility. The importance of stretching before and after any kind of physical activity cannot be overemphasised. Newcomers (especially men) often classify stretching as 'cissy stuff' and not worth considering as worthwhile exercise. It is, however, a very good way of keeping injury at bay. There is nothing more frustrating than starting on an exercise programme, only to be injured with pulled muscles after a couple of days. Remember – stretch first, then exercise and don't be too ambitious at first. Build up gradually.

**Guidelines for stretching**

- Relax and gently ease into the stretch.
- When you feel mild tension in the muscle being stretched, hold the stretch.
- Do not bounce in an attempt to reach further.
- Slowly ease out of the stretch.
- Stretch both sides of the body.
- Do not hold your breath.

A small range of stretches is described in **Table 6.2**. These focus on stretching the leg muscles which are the ones most likely to suffer injury if used more energetically than usual.

| STRETCH | POINTS OF TECHNIQUE |
|---|---|
| Lower leg (calf) | Step one foot in front of the other and bend the front knee so that the knee is directly over the ankle, and both feet are facing forwards. With hands on hips, press the back heel into the floor. |
| Back of thigh (hamstring) | Step one foot in front of the other. Bend the back leg and push your bottom upwards and outwards. Keep back straight, look forward and place hands on your bent leg (supporting leg). To increase the stretch, lift the toe of the straight leg towards the ceiling. Repeat using opposite leg. |
| Front of thigh (quadriceps) | Feet hip-width apart and soft knees. Bend one knee holding the foot with the same arm and bring it inwards towards your bottom. To increase the stretch gently push the hips forwards. You can use either a partner or a wall for support. Repeat using opposite leg. |
| Inner thigh (adductors) | Stand facing forwards with feet apart. Turn one toe out and bend the same knee so it is directly over the ankle. Place hands on hips. To increase the stretch, start with feet further apart. Repeat on other side. |

**Table 6.2: Stretching the leg muscles.**

**Walking and jogging**

You should start with brisk walking and try to increase this gradually to 45 minutes. You should then introduce slow jogging, eg walk 100m, jog 100m. The amount of jogging should gradually be increased with the aim of jogging for 20 – 40 minutes continuously. When you can jog for at least 20 minutes you may still want to do some brisk walking for variety as a separate session.

*Sample programme over 10 weeks*

| W | Session 1 | Session 2 | Session 3 |
|---|---|---|---|
| 1 | 20min brisk walk | 25min brisk walk | 25min brisk walk |
| 2 | 25min brisk walk | 25min brisk walk | 25min brisk walk |
| 3 | 25min brisk walk | 30min brisk walk | 30min brisk walk |
| 4 | 30min brisk walk | 30min brisk walk | 35min brisk walk |
| 5 | 30min brisk walk | 35min brisk walk | 40min brisk walk |
| 6 | 20min alternate walk jog | 40min brisk walk | 20min alternate walk jog |
| 7 | 20min alternate walk jog | 45min brisk walk | 20min alternate walk jog |
| 8 | 10min jog, 20min brisk walk | 20min alternate walk jog | 10min jog, 20min brisk walk |
| 9 | 15min jog, 15min brisk walk | 25min alternate walk jog | 15min jog, 15min brisk walk |
| 10 | 20min jog, 15min brisk walk | 25min alternate walk jog | 20min jog, 20min brisk walk |

You should gradually build up session three so you are exercising for an hour with a combination of brisk walking and jogging.

**MAP-READING SKILLS**

Those who come into ARDF from a radio background often find that their map reading lets them down at first. The orienteering maps used for ARDF competitions have some apparently idiosyncratic features as far as the newcomer is concerned. For example, someone brought up to read Ordnance Survey maps will have to adjust to the fact that on orienteering maps, white areas are runnable forest and not open land.

Both map-reading skills and general fitness can be improved by joining the local orienteering club. Most clubs run an event every month and, in most areas, there is an event within drivable range each Sunday. Participation in these soon improves personal performance at ARDF.

**RANGE ESTIMATION**

Time spent in practice developing adequate range estimation will be well rewarded in competition. Beginners often stop too soon and wait for the next transmission. If they had only come up with a better estimate of the distance to the transmitter, then they would make a better estimate of how far they need to run before stopping to take the next transmission from the fox they are hunting.

To do anything useful to improve range estimation, an ARDF transmitter is required. These transmit the correct power with a typical ARDF antenna and will therefore give representative signal strengths. A local ARDF group will probably have both 3.5MHz and 144MHz transmitters available and can stage a training session. Failing that, use the transmitters in a competition. After completing the course, listen to the signals being received at the finish from the competition area and note their strengths.

The results need to be recorded in some way. The easiest method is to mark range estimates on either the volume control or the 'squeaker' control in the case of receivers with an audio S-meter. The reader will recall that the hearing sense should be used for the purpose of range estimation, because using the eyes to look at an S-meter slows up progress towards the fox. Most competitors set the volume control to achieve a 'comfortable' volume. The volume control can then be calibrated with distance.

Clearly, this is not a precise technique, but efforts should be made to 'calibrate' the receiver with perhaps three key distance markers. The technique, at best, is capable of distinguishing a transmitter at 100m distance from one at 200m and then one at 200m from one at 400m.

Perhaps the most important distance marker is 100m. If a transmitter comes on the air at this distance (or less) then it should be easy to run down the bearing and locate it while it is still transmitting. Markers on the scale for 400m and 1000m are also useful. Transmitters between these distances should be easy to find on the next transmission. If the estimate of distance is reasonable, it should be possible for the competitor to reach a position close to the fox in readiness for the start of the next transmission in four minutes' time. Finally, if the range comes up as significantly over 1000m, then it is a case of 'keep going'.

Both frequency bands have peculiarities in respect of range estimation that need to be considered.

144MHz is prone to multi-path propagation and screening. Distance estimates need to be considered in relation to the terrain. The planner will often try to confuse competitors into making inaccurate estimates of range by placing close-by transmitters in sites which have some local screening and placing transmitters which are far away, on the top of hills or with ground dropping away from them in the direction of the start. Aware of this, the wise competitor will check the nature of the terrain as shown on the map to decide the 'weight' to be given to the range information.

In carrying out range calibration for 144MHz, this should be done in a flat forest with little or no multi-path propagation.

The difficulty with 3.5MHz is to decide whether to conduct the range estimation with the whip antenna connected or not. It is a bit of a fag to press the button and rotate the receiver because, when this is done, the 'comfortable volume' is likely to be at a lower volume control setting than was being used to check the null. Hence, the competitor has to fiddle with the volume control as well as pressing the button and rotating the receiver. It is better to use the broad maximum at right angles to the null when estimating range. The volume control will be set close to a comfortable volume in this direction in any case.

The procedure then becomes as follows.

- When the wanted transmitter 'fires up', set the volume to a comfortable level.
- Find the null and, if appropriate, plot the bearing.
- Turn the receiver 90° to the null, verify the volume is 'comfortable' and do not press the sense button. Then read the approximate range from the calibration of the volume control.

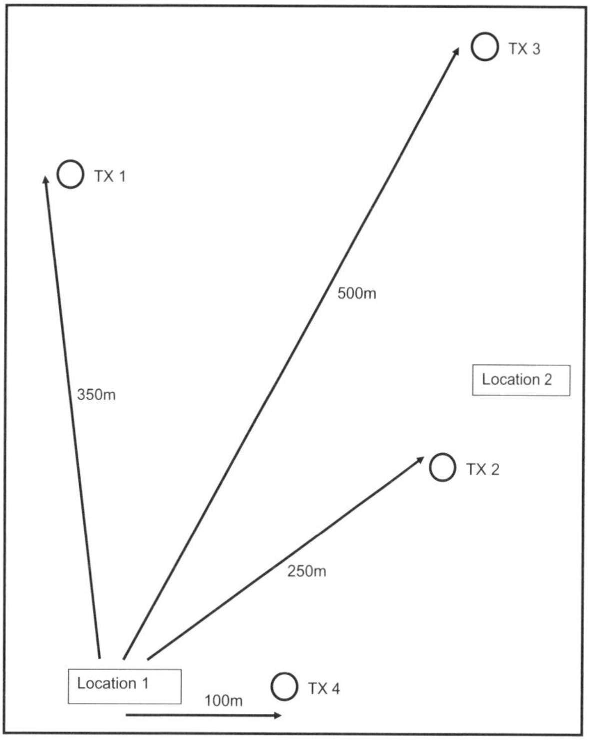

**Fig 6.1: Range estimation exercise. Participants estimate the distance to the four transmitters and then move to location 2 to repeat the exercise. When the locations of the transmitters are revealed to them, they can assess the accuracy of their range estimation.**

## TRAINING EXERCISE 1

Five transmitters are deployed at distances between 100m and 1200m from the point where the exercise is carried out (**Fig 6.1**). Participants have to estimate the distance to each transmitter. This can be combined with taking and plotting the bearing to each transmitter.

Then, the group moves to a second location and even a third to repeat the procedure. Finally, individuals check their results against the locations of the transmitters as revealed to them at the end of the exercise. Look for systematic errors in your results. Modify the range scale on the volume control as necessary.

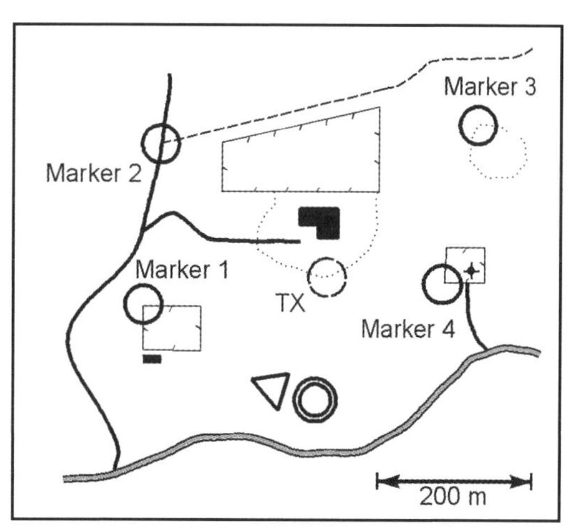


**Fig 6.2: Exercise to practise the taking and plotting of bearings to a hidden transmitter.**

**Taking and plotting bearings**

Taking and plotting a bearing is a fundamental technique and the methods were described in Chapter 3. Here are details of a training exercise to assess the ability of participants in this regard.

**TRAINING EXERCISE 2**

Using a mapped area with a map to a scale between 1:5,000 and 1:10,000, set out four to six control markers roughly positioned on a circle of about 600m diameter (**Fig 6.2**). Participants have to visit all these locations (and punch a control card if the organiser so ordains) but, at each control location, they are required to take and plot a bearing to a transmitter located somewhere inside the circle. The distance to be covered is about 2km, which is not excessive for most people.

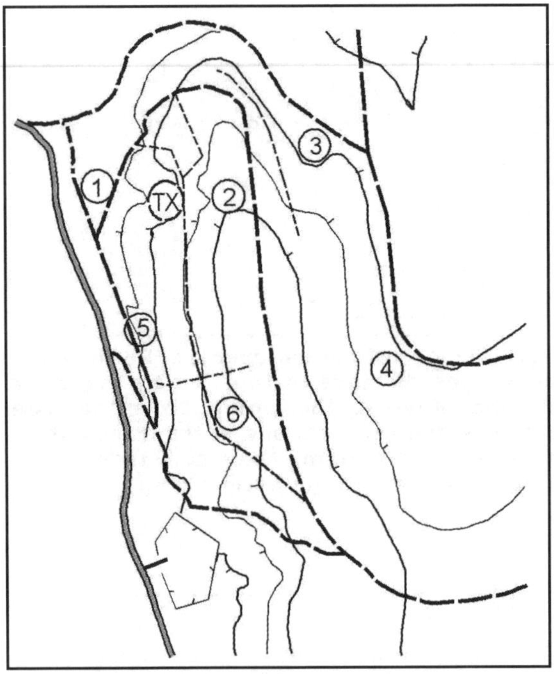

At the conclusion, each participant, on being told the correct location of the transmitter, can assess the accuracy of his / her bearings. Any systematic error, such as all bearings error to the left, can be assessed in detail. Extensions to the exercise might be as follows.

- Go round a second time just visiting the control sites and taking no bearings. The difference in time to the original circuit represents the time needed to take the bearings.
- Move the hidden transmitter and repeat the exercise to check if an improvement in performance can be obtained.

**Fig 6.3: The transmitter is placed in a re-entrant and then bearings are taken from points all around to assess their accuracy. (The tags on the contour lines are on the downhill side.) Take bearings looking up and down the re-entrant (1 & 2); from behind the hill (3 & 4); along the hillside from a position low down (5); and a position high up (6).**

**Mastering multi-path propagation**

Direction-finding using the 144MHz band in hilly terrain is bedevilled by multi-path propagation. Signals arriving from directions other than the direction to the hidden transmitter are confusing and misleading. The experienced competitor develops a link between the terrain as represented on the map and the likelihood of experiencing multi-path propagation.

It is possible to contribute to the gaining of this 'experience' in a training situation rather than suffering disappointing results in competition when hilly terrain is used.

**TRAINING EXERCISE 3**

This requires a piece of suitable terrain not dissimilar to that shown on the map in **Fig 6.3**. A country park or other open area may have

sufficiently detailed contouring and a significant slope with re-entrants and spurs. Better still, this might be on one side of a hill so that the effect of finding and taking bearings of a transmitter from the reverse slope can be absorbed.

The exercise can be conducted solo and the only requirement by way of equipment not already owned is the use of a 144MHz DF transmitter and associated antenna. It is important to mimic the transmitters used in competitions in order to give realistic signal strengths. A map of the area will also be required and, in this regard, the majority of country parks in the UK has been mapped for orienteering purposes at some time or another.

Place the transmitter in a re-entrant on one side of the hill and set it up as a beacon transmitting MO continuously. Then move around the area taking and plotting bearings to the transmitter noting the locations where the bearings are reasonably accurate and more importantly, locations where the bearings are significantly in error. Learn from these results so that it is possible to recognise situations in competition where it is likely that the bearings will be erroneous and then to move to a better location and discard the poor-quality information.

The final 400m

The last 400m of a hunt for a transmitter can be a critical time. Get it right and the transmitter is found during the 60 seconds it is transmitting or shortly after it ceases transmission. Get it wrong and the competitor has to wait for four minutes before it transmits again. That four minutes can be the difference between a high place in the results and a lowly one.

In the last 400m of the hunt, the signal strength of the transmitter starts to alter noticeably as the competitor changes position and the bearing to the transmitter can also be changing rapidly. The competitor must react quickly to these changes and interpret them correctly to locate the transmitter.

**TRAINING EXERCISE 4**
This exercise requires a partner, a suitable area (eg a country park), a couple of orienteering maps of the area and a small ARDF transmitter that can be put on air within 60 seconds of finding a suitable location.

The first member of the duo sets off into the area with the transmitter. This is placed somewhere about 400m from the start point and is then 'keyed up' for one minute in every five. The second person, on hearing the transmitter come on the air, then attempts to locate it as a result of the first one minute transmission alone.

The roles are then reversed and the second member of the duo takes the transmitter and places it in a location about 400m further distant for the first member to find. The process can be continued for as long as desired.

CHAPTER 7

# THE INTERNATIONAL PERSPECTIVE

ARDF using the IARU rules can be distinguished from the other forms of direction-finding that operate in the UK, in that it has a strong and well-developed international dimension. It forms a branch of 'Sport Radio' where radio amateurs with a competitive streak seek to test their ability and skills against other like-minded people in friendly competition. The most obvious manifestation of this is radio contesting, but there are lesser-known areas of competition such as high speed telegraphy.

**Competitors at a World Championship check the results displayed at the finish.**

There are some who question the concept of 'Sport Radio' since they feel that some kind of physical ability needs to be demonstrated in a sport. ARDF clearly satisfies this kind of objection since it involves moving on foot through woods and open terrain as fast as possible.

Chapter 1 outlined the history of ARDF international competition and, in this chapter, there is an overview of the current international structures and competitions.

## INTERNATIONAL COMPETITION

Very few of us ever get the opportunity to represent our country at some kind of sporting activity and simply watch others on television. There is a world of difference between participating in a local event in the UK and

**Teams at the opening ceremony of the 2004 World Championships.**

**A Ukrainian competitor 'clears' his 'dibber' at the pre-start.**

being a part of a full international championship. There is the process of accreditation followed by an opening ceremony with all the teams parading with their national flags. Speeches and folk dancing seem to be a staple component of these ceremonies. Following each of the two competitions there will be a medal presentation ceremony with a podium, flowers and the treasured medal for the best three in the class.

Aside from the ceremonial aspects, participants spend five days in the company of fellow enthusiasts from many countries and there is the opportunity to look at the equipment they use, talk to them about their race and learn from the best people in the world. In Region 1 we are lucky in that the Region 1 Championships in odd numbered years will see many of the same people on the podium who will be there in the next World Championships, such is the strength of ARDF in Region 1. The last evening of championships will see a 'hamfest', a social gathering for all the participants and officials. This is an opportunity to cement friendships and catch up with old friends. It is the custom to exchange small gifts.

The national society of each member country is invited to send a team comprising three competitors in each of the five men's and four women's classes. Hence a full team is 27 persons. In the team competitions, the best two members count to decide the winner. Team selection arrangements vary enormously from country to country depending on the level of interest in ARDF in each nation. Some countries choose the date of their national championship to be able to use the event as a selection race for the world or regional event later in the year.

**Watching the finish at the 2004 World Championships.**

The venue is the choice of the host nation and is decided by the availability of suitable forests and accommodation as well as accessibility for those flying in, perhaps from the other side of the world. The costs of the event are borne by the participants and their national societies. Some sponsorship is usually forthcoming, but this is used to offset the direct costs of staging the event.

**The flower ceremony at the 2004 World Championships.**

There is a lot of prestige attached to the privilege of representing one's country. The standards achieved by the winning individuals and teams can be a bit intimidating but the athletic prowess and technical ability of the winners has to be admired. It is always interesting to see how you, as an individual, shape up alongside the best in the world. In some cases the very best are not totally out of sight either, and this can spur one to try to improve even more in the future. The age group aspect of the structure of the sport is a strong incentive to do well in the year one moves up into the next age category.

**ARDF AND THE IARU**

All international sports have a governing body and, in the case of ARDF, this is the International Amateur Radio Union (IARU). For a sport that lies midway between amateur radio and orienteering, it is not necessarily clear why this became the case. A look back at history tells us that ARDF grew from early post-war radio direction finding and the Nordic influence helped to direct this down the route of a pedestrian sport not involving the use of motor vehicles. This background coupled with the need to have transmitting licences for the hidden transmitters determined that it is the IARU that represents the sport at international level.

The IARU is a federation of national radio societies that was created in 1925 and is divided into three regions. Put simply, Region 1 covers Europe, Africa, the northern parts of Asia and the Middle East, Region 2 is North and South America whilst Region 3 takes care of South Asia and Oceania.

The IARU ARDF Working Group is the body which establishes the rules and sponsors the international competitions. It was established in 1978 and this heralded the start of the international perspective to ARDF that we see today. Applications are welcomed by the Working Group, from member Societies, to stage major international competitions. The World Championships is a bi-annual event and IARU Regions stage their regional events during the intervening years. At present, the World Championships take place in even-numbered years. These events usually take place in early- or mid-September when the major holiday period is over but before the university students return to campus.

**THE ORGANISATION OF MAJOR CHAMPIONSHIPS**

The ARDF Working Group accredits international-class referees being persons who are experienced participants in national and international events and are proposed as Referees by their National Societies and accepted as such at a formal meeting of the Working Group.

For each international competition, there is an international jury made up as follows.

- Chairman
- Secretary
- Start referee
- Finish referee
- Referees in the competition area
- Siting referee
- Technical director.

The siting referee, the referees in the competition area, and at the start and finish, have to be international-class referees. There is no requirement that the secretary or the technical director be international referees.

It is often the practice that each transmitter in the competition area is under the observation of one of the competition area referees as well as an 'operator'. The latter is able to substitute equipment in the event of a technical failure. It it also possible for the referees in the competition area to be roving rather than fixed in position.

**Queuing at the finish to download data from the dibbers carried by the competitors. In the centre is Josef Tuttmann of the Netherlands.**

The siting referee has the very challenging task of deciding the locations of the transmitters. With only five transmitters to be deployed to cater for the full range of abilities from M21 to W50, this is a near-impossible task. The M21s, M19s and M40s need a reasonably long course to offer the sort of challenge expected at the highest level of competition. Conversely, the older age groups of W50 and M60 cannot be expected to cover long distances within the time limit set down by the International Jury.

From 2007, the Working Group has agreed to a system of grouped competition to alleviate this problem. On a given day of the competition, transmitters for *both* frequency bands will be deployed. Those for 3.5MHz will be located within a reasonably small area suitable for the older age classes whilst those for 144MHz will be much further apart and will be hunted by the youngest classes. On the other day of competition, the deployment is reversed with the younger age classes hunting wide spaced transmitters on 3.5MHz while the older age classes have to find 144MHz transmitters which are much closer together than they were previously. This change will make it easier for the siting referee to meet the requirements of the wide range of age classes so that all groups experience an appropriate standard of international competition.

**Part of the RSGB team at the 2006 World Championships. L – R: Bob Titterington, G3ORY, Robert Vickers, G3ORI, and David Heale, G6HGE. (Photo: David Williams, M3WDD)**

Notwithstanding these difficulties, the siting referees at international competitions have had some striking successes in the past, in tempting competitors to take non-optimal routes to the hidden transmitters.

The start referee is responsible for the arrangements at the start. Competitors have to be corralled separately from their receivers. A warm-up area must be provided. The start itself must comply with the rules regarding the non-visibility of the competition area and the rules for map issue and start facilities have to be observed. The provision of some limited cover in the event of inclement weather, clothing transfer to the finish area and the prevention of cheating by the use of mobile phones etc add to the problems faced by the start referee.

The finish referee has to establish the facilities at the finish. The advent of electronic punching facilities from commercial firms has made the provision of up to date results much easier than it was in earlier years. It is normal to find computer monitors displaying up-to-the-minute results for each of the courses and there have been events where the results are uploaded minute-by-minute to a website so they are instantly available world-wide.

As soon as a competitor has punched at the finish, he or she 'downloads' the data from the electronic chip carried and a computer system then updates the result table virtually instantaneously.

It is the international jury as a whole that has to fix the time limit for the event. In international competition this may well be in excess of 120 minutes, particularly if all age classes are using the same set of five transmitters. When the race is over, the results have to be approved by the International Jury. They also deal with protests where voting is used to reach the decision. The technical director and secretary do not have votes.

**PROGRAMME FOR AN INTERNATIONAL COMPETITION**

The programme for these events has become fixed and the following format is used

**Tuesday** Competitors arrive and are accredited. Entry and accommodation fees have to be paid and accreditation shows that all is in order.

**Wednesday** Half of the day is given over to a training session. 3.5MHz and 144MHz transmitters are deployed in a small area, normally within walking distance of the accommodation. It is an opportunity for competitors to test that their receivers have survived the worst that airport baggage handlers can do and also to assess the strength of the signals being radiated by the transmitters provided by the host nation. Any personal form of range estimation can be reviewed in the light of the results obtained.

In the afternoon, the opening ceremony is held, when the competitors of the participating nations parade with national flags. As noted earlier, this ceremony usually includes speeches and folk dancing.

There will also be an ARDF Working Group meeting, as well as a briefing session for team captains.

**Thursday** Day one of competition. In earlier years, it was the norm for only one frequency band to be in use, but the conflicts facing the siting referee mean that, in future, both bands are likely to be used, with the older age categories competing over courses planned around a set of transmitters which are not too far apart on one band. The other frequency band has widely spaced transmitters for the younger age categories.

In the evening there will be a medal presentation.

**Friday** Free day with an organised trip to a nearby site of tourist interest.

**Saturday** Day two of competition. In the evening, the medal ceremony takes place followed by a 'Hamfest' at which competitors are able to socialise, exchange small gifts and make friends who share the same interests from across the world.

**Sunday** Journey home

**IARU YOUTH CHAMPIONSHIPS**

There is provision within the rules of ARDF for youth championships. These are held within IARU regions and there is no world-wide event. The Championships cater for competitors aged 15 or younger in the classes M15 and W15. Each National Society may enter two teams of three in each category.

The M15 category hunts three to five transmitters over a course of length four to six kilometres while the W15 category hunts three to four transmitters over a course length of three to five kilometres.

**The US team pose for a photograph at the World Championships in 2004.**

The rule excluding transmitters from a zone within 750m of the start is relaxed to 500m for this championship, although transmitters must still be at least 400m from each other and from the finish.

Only Region 1 of the IARU regularly stages a Youth Championship.

These events are run solely at the discretion of the National Society concerned. They use the rules adopted by the IARU ARDF Working Group, but the formats can be very different depending, mostly, on the level of support within that country.

There are generally two 'models' for such an event. Some societies have a weekend of competition with a 144MHz event on the Saturday and a 3.5MHz event on the Sunday, often with some sort of 'Hamfest' on the Saturday evening.

Others try to compress both competitions into one day and sometimes even require all competitors to hunt all the transmitters in both competitions. This can be quite a burden for the older age groups. More often, competitors in each age group hunt the number of transmitters normally specified for that age group.

CHAPTER 8

# EQUIPMENT

## Overview

Many amateurs thrive on experimentation so, for those, it will be good news to hear that ARDF offers unique opportunities for construction and development, whether for the receivers carried or the transmitters deployed. The analogy with QRP construction is strong. Simple home-built receivers and QRP transmitters are routinely as effective as any available commercial equivalent.

Apart from the reward of building and using your own receiver, there is plenty of scope for the experimenter to refine both receiver and antenna components to provide a more effective integrated DF unit. QRP transmitters are a lot easier to build these days, using modern components, even at VHF. Transmitters are nearly always combined with a microprocessor chip to control delayed operation and the generation of an identifying callsign in Morse code. Amateurs who like experimenting with software code may enjoy working with ARDF transmitters.

Like other spheres of amateur radio, the reasons for interest in ARDF will be varied. For those who come from an orienteering background, the receiver may just be a necessary tool to compete. Many amateurs, however, will use ARDF to prove their latest technology implementation and at the same time, enjoy radio foxhunting, in attractive woodland surroundings. The competitive element of ARDF is often trying to outdo their amateur radio peers.

Equipment development may aim to improve the users' competitive standing, but finding out if such changes actually achieve this, will be half the fun. By way of example, many traditional 144MHz DF receivers use a free-running local oscillator (LO) which is perfectly adequate, when working with amplitude-modulated signals. Nevertheless, experimenting with a phased-locked loop (PLL) voltage-controlled oscillator (VCO) is a popular project for ARDF enthusiasts

80m receivers have a simple sensing method, which typically provides a front-to-back signal ratio of just 7dB. Enough for most, but others will be happy to spend time and effort to improve this figure, to give more reliable results over less favourable terrain.

The disappearance of many of the radio integrated circuits (ICs) that were so common in the 1970s has led to a new raft of problems. For example, finding an amplitude-modulation (AM) intermediate-frequency amplifier chip is well nigh impossible today. As a result, some of the older designs of ARDF receiver are no longer viable and, as demonstrated in some of the designs described in this chapter, considerable ingenuity has been used to utilise ICs for purposes which their designers probably never envisaged. This has been the thrust of most of the circuits developed recently.

The use of AM for 144MHz is unique in the amateur radio arena. There is enormous scope for amateurs to dust off some off some of the older AM technology and re-apply it with modern components and ideas. Simple IF amplifiers based on MMIC or OpAmp devices are perfectly viable, especially as the extra complexity of automatic gain control (AGC) is not required.

Receiver design is very important, because every competitor needs one. In a big international competition, receivers of every shape, size, complexity and sometimes effectiveness are to be seen. There is, however, a common thread in their design. Since ARDF is conducted on foot, the receivers need to be light and easily portable. Since the transmitter powers are specified in the IARU rules, the signals will be reasonably strong over the competition area, and it is not necessary to look for low-noise high-sensitivity, receiver performance.

ARDF involves deciding the direction of arrival of signals based on variations in signal strength as the antenna is rotated. Consequently, the best receiver architecture is an AM receiver without AGC. This arrangement will make the changes of signal strength most apparent and enable a better decision of the direction of arrival to be made.

It is possible to use a frequency-modulation (FM) receiver and many people do, especially in the first few competitions they enter. However, the limiter in an FM receiver and the likelihood that any AGC cannot be disabled, both make the signal amplitude changes less apparent, especially if judged on the loudness of the audio output alone.

A requirement for ascertaining the direction of the fox (born of practical experience), which has already been mentioned in this book, is the importance of relying on the sense of hearing when using the receiver. To do well in a competition, the receiver has to be used on the move and, in these circumstances, eyesight is occupied in avoiding running or walking into anything. The ears have to be used to judge signal strength rather than using the eyes to look at an S-meter. This reinforces the need for a receiver design in which signal amplitude changes at the antenna, result in closely-matching changes in audio output. It is the AM receiver without AGC that delivers this.

At VHF the IARU rules are based on technology which was current several decades ago. The modulation on the 144MHz transmissions is modulated carrier wave (MCW), ie a keyed audio tone, amplitude-modulated on the carrier, conveying the identification data for each transmitter. There is some variation between countries regarding whether the carrier is continuous, with keyed audio tone OR whether both carrier and audio tone are keyed together, but both methods are supported in the IARU Region 1 rules. MCW was used in those days mainly because the frequency stability of the receiver LO was much poorer compared to a modern single sideband (SSB) receiver. MCW allows the identifying tone to be demodulated to give the desired audio frequency, even if the receiver LO drifts a little.

As a result, receiver design can be fairly simple and perfectly acceptable results obtained. In a single-conversion superhet for 144MHz use, the LO will run at a VHF frequency but, by using a wide IF filter, the stability of the LO is much less critical. A IF filter with a -3dB bandwidth of 20kHz is probably a practical minimum, for the user to be unaware of slight frequency drift. IF filters of between 30 to 50kHz are ideal. Adjacent channel interference is rarely an issue. Compare this to an SSB receiver where any LO drift directly changes the recovered tone frequency (assuming a keyed carrier as used for ARDF on 3.5MHz). Now a drift of a mere 50Hz causes a noticeable change of tone frequency.

These factors mean that it is possible to build a perfectly-acceptable 144MHz ARDF receiver using just two ICs. This also applies to 3.5MHz where the frequency stability problems are not as onerous as at 144MHz and the IARU rules specify the use of keyed carrier to convey the transmitter identification information. Hence, the ARDF receiver

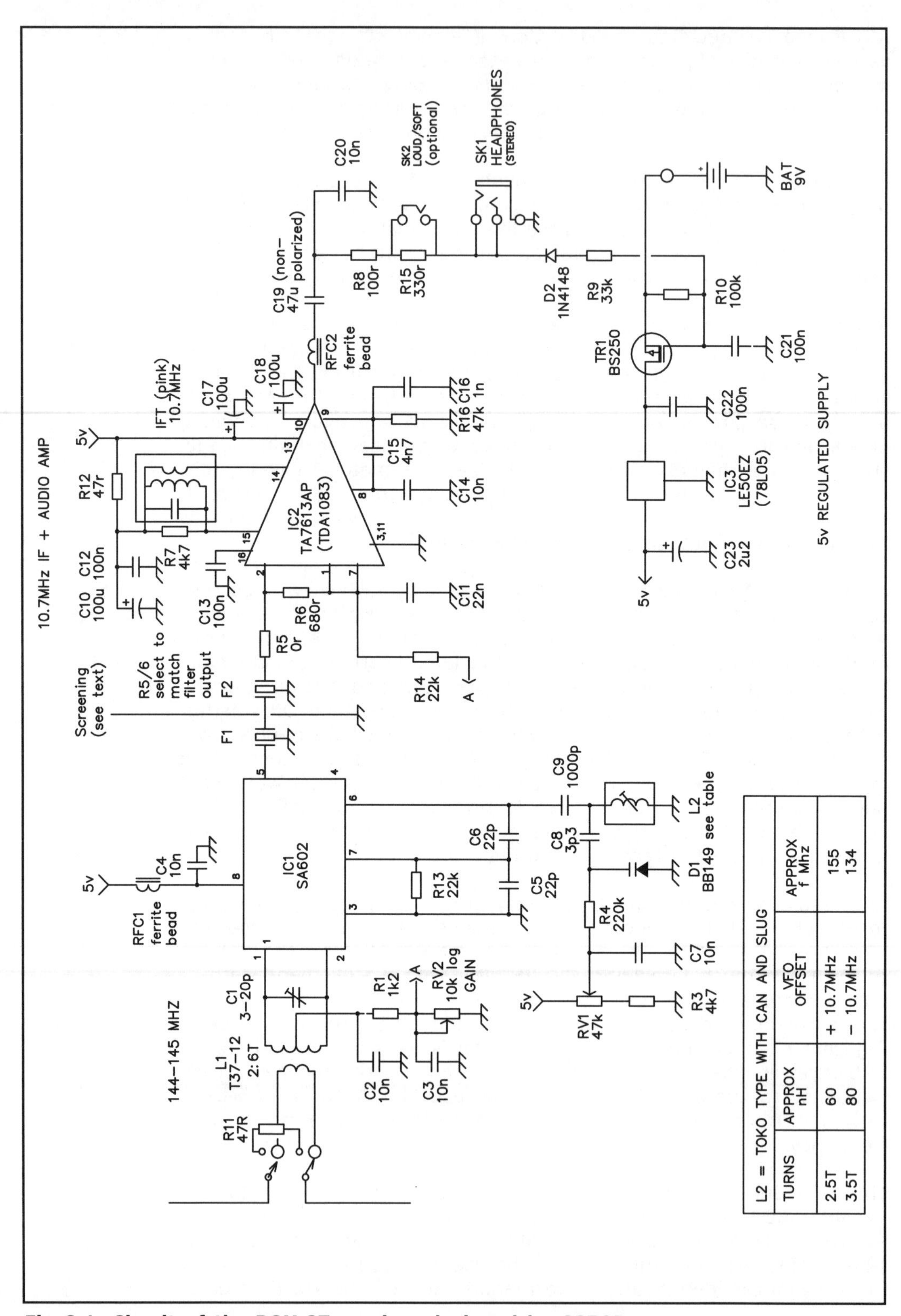


**Fig 8.1: Circuit of the ROX-2T receiver designed by G3ZOI.**

makes a wonderful group or club project since the resulting receiver is not a rather poorly-performing artefact, which is soon discarded, but a completely effective receiver for participating in ARDF competitions. There is enormous pleasure and a great sense of achievement to be had in using a self-constructed piece of equipment to take part in an amateur radio competition.

ARDF is one of the last bastions of home construction. The equipment is sufficiently simple for it to be practical to embark on 'rolling your own' whether it be from a kit or from a circuit diagram alone.

## RECEIVERS

### A SIMPLE 144MHz RECEIVER: ROX-2T

The incentive for this project came after the designer, Dave Deane, G3ZOI, dismantled a rusty AM/FM shower radio and discovered an interesting-looking 18-pin chip marked 'TA7613AP'. A quick Google search revealed this chip to be a variant of the TDA1083. The TA7613AP is a one-chip AM/FM radio, complete with audio amplifier.

The TDA1083 chip is a popular choice for ARDF designs from Eastern Europe, notably from OK2BKN and YU7AM, and a degree of similarity is acknowledged in the ROX-2T.

The ROX-2T is a single-conversion superhet receiver (**Fig 8.1**) with a free-running local oscillator and a 10.7MHz intermediate frequency (IF). In similar designs, an additional RF amplifier provides gain control, but is absent in the ROX-2 by using a novel way of controlling the gain of the SA602 mixer (as found in the Elecraft K1 transceiver). The bias in the SA602 is externally varied through a centre tap on the secondary of the antenna coil. With this arrangement it is important either to feed the SA602 directly with the stabilised supply voltage to the chip, or a through an RFC, to minimise pulling of the LO.

The SA602 provides one of the best stable LOs around and includes temperature compensation, so the device itself does not contribute significant drift. Careful design of the VFO is still required to limit drift caused by ambient temperature changes. NPO capacitors are used where possible. Using a low-value inductor for the LO coil and relatively high-value capacitors, reduces the drift caused by stray capacitance. In this respect, opting for a LO running 10.7MHz below 144MHz may provide improved stability, even if there is a greater risk of image interference from airband frequencies.

The antenna input to the SA602 uses the balanced input to the chip and the receiver is designed to be mounted at the feed-point of the antenna with short leads in a completely balanced arrangement. A dust iron toroid is used to transform the low antenna impedance (approximately 25Ω) to a higher value. With the simple input coil arrangement shown, the impedance matching is not optimum; however, the resonant secondary winding provides selectivity, some image reduction and more than adequate sensitivity at 144MHz. One very important aspect to note is that both halves of the secondary winding must be wound in the same sense.

Selectivity at the IF is provided by 10.7MHz ceramic filters with a -3dB bandwidth of 20kHz. Unfortunately, the choice of narrow-band filters at 10.7MHz is becoming increasingly limited. The stop-band characteristic of the Token filters used is quite poor, so two are cascaded to improve the attenuation of adjacent-channel signals. This factor is probably the most difficult parameter to achieve with simple 2m receivers. In the deforested UK, the restricted size of available woodland means ARDF courses are normally circular, with the home beacon being located close to the start. The home-beacon-to-fox frequency spacing will typically be 200kHz or less, in which case the home beacon will be extremely strong at the start and can cause difficulties when initially trying to detect signals from the more distant foxes.

The current design shows a small grounded screen made of brass shim or copper foil, placed between the two filters on the component side of the PCB, to restrict signal bypass and the resultant degradation of the stop-band rejection. Further attention to isolating the mixer from the IF amplifier would probably be worthwhile.

Following the filters, the signal then enters the TA7613AP / TDA1083, configured as a 10.7MHz AM IF amplifier with demodulation and AF amplification, which delivers up to 300mW into a pair of headphones.

The gain of the IF amplifier is modified simultaneously with that of the SA602 by using the same gain control rail to pull down the voltage on pins 1 and 7. Further signal attention is achieved by using a 'brute force' method – breaking the antenna connection to the SA602 input with a DPDT toggle switch. The signal path to the receiver is then only via stray capacitance within the switch, allowing the user to direction-find right up to ARDF transmitter antenna. The attenuator switch stops very strong signals saturating the ROX-2T and provides an excellent range indicator for the closeness of a transmitter, ie if the signal is strong even with the switch open, the user is within visual distance.

The BS250 (TR1) acts as a battery switch controlled by a tiny DC current flowing through the headphones when they are plugged in. Many designs use the plug of mono headphones inserted into a stereo socket to provide a battery switch with one side grounded. However, mono headphones are less readily available and are more expensive than stereo ones.

The resulting ARDF receiver is just about as simple as a superhet can get with get two RF ICs, and no surface-mount components, apart from the varicap diode. It is easy to build with the preponderance of leaded components. As well as being simple to build, the receiver is simple to use and a newcomer will easily and quickly grasp the functions of the two controls and the need to swing the antenna for maximum signal.

The receiver is built in an RFI (radio frequency interference)-coated ABS box and can conveniently be mounted directly behind the feed-point of the three-element tape beam described later in this book. The all-in weight of the receiver and the antenna is an unbeatable, lightweight 390g.

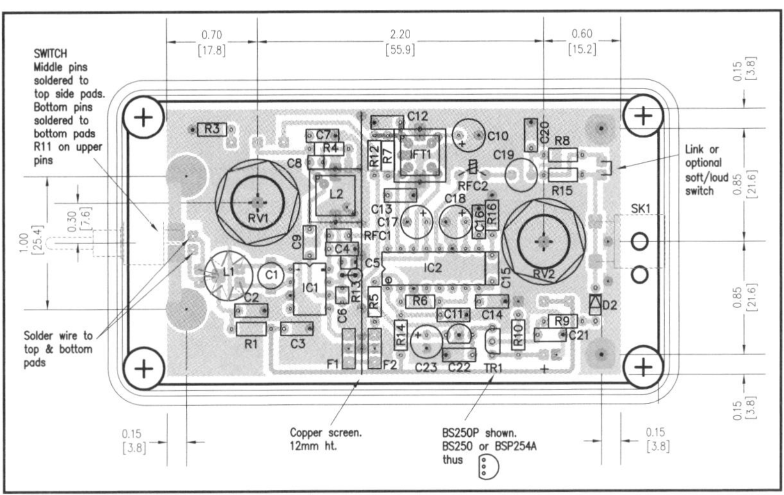


**Fig 8.2: Component overlay for the ROX-2T.**

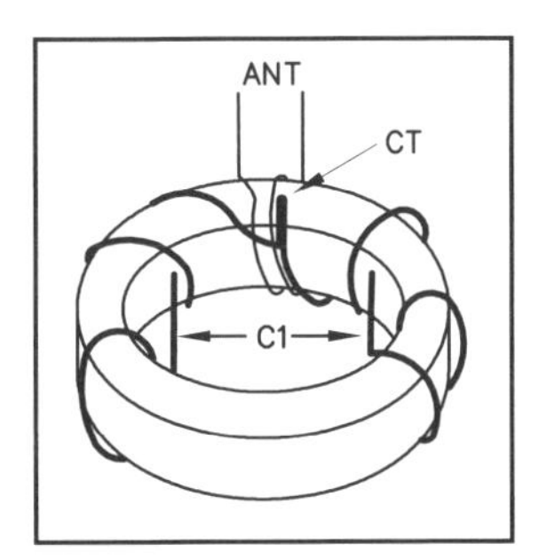


**Fig 8.3: Winding detail for the toroidal input transformer.**

The ROX-2T makes an ideal club or group construction project. The completed receiver can then be used to participate in ARDF competitions. **Figs 8.2**, **8.3** and **8.4** provide the constructional details.

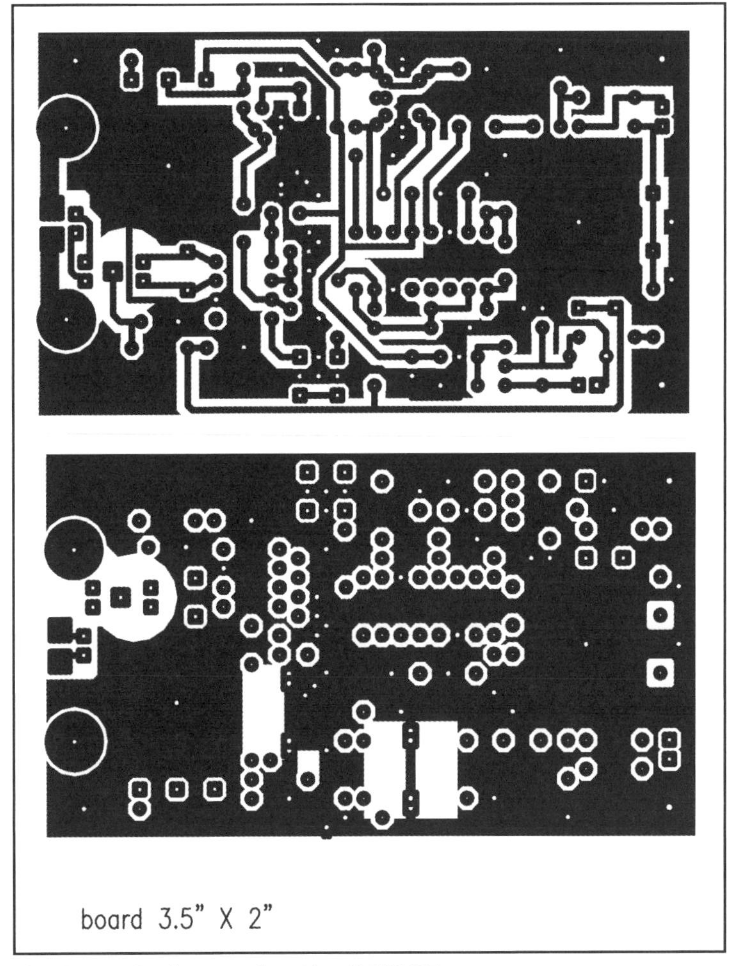


**Fig 8.4: PCB layout for the ROX-2T.**

**ANTENNA FOR 144MHz DIRECTION-FINDING**

Of course, any 144MHz Yagi antenna can be used. However, for best results there are several factors to consider.

- The greatest impediment to obtaining good bearings on 144MHz is the presence of multi-path propagation. The signals from the transmitter can arrive at a variety of angles having been reflected from hillsides, the edge of woodland or man-made objects. Discriminating between these reflected signals and the direct signal from the transmitter is always problematic and in some cases impossible. The situation is eased if the receiving antenna has a particularly 'clean' pattern, ie the side and back lobes of the Yagi are well suppressed by virtue of the design. This is a much more important aspect of the design than forward gain. Commercial Yagis, built for a wider market, will always focus on gain since this is a strong selling point.
- 'Yagis always work'. This statement reflects the truth that if an element about five percent longer than a half-wave dipole is placed behind such a dipole and a third element about five percent shorter than a half wave is placed in front, then the aerial will 'work' in that forward gain is observed. Getting it to work well is much harder and never more so than when trying to suppress side and back lobes of the polar diagram. Computer modelling has come to the fore in the last twenty years and it is now possible to assess a given design by using a *NEC*-based computer programme.
- It is easier to get good sidelobe suppression with a three-element Yagi than it is with two elements. The latter can give adequate performance, but any two-element solution needs to be assessed very carefully to decide if the suppression is adequate. Having said that three elements are good, why not add a few more? The answer to this question lies in the area of practicability. One of the lessons of the computer modelling results is that overpopulating a boom of a given length with elements, leads to no improvement in performance. Focus on the longest boom that can be conveniently carried and used on foot in wooded areas; anything over three feet (900mm) in length is starting to be unwieldy. For this boom length, no more than three elements are appropriate
- Finally, it is necessary to consider the mechanical arrangements. Beams based around plastic components, usually water pipe or waste pipe and the associated fittings, are very light in weight and the components are cheap. The elements are made, to advantage, from pieces of steel tape measure. If high gain was the priority, using steel for the elements would not be countenanced. However, side- and back-lobe suppression is king and using lossy elements is not a disadvantage. The huge advantage of using steel tape measure for the elements is that they will bend readily when moving through close spaced trees or scrub but then instantly spring back to their original shape in clearer terrain

## A PRACTICAL YAGI ANTENNA FOR 144MHz ARDF

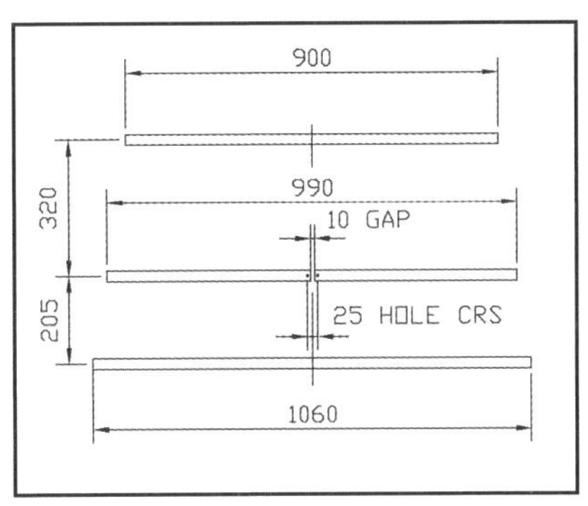


**Fig 8.5: Diagram of the antenna.**

This design (**Fig 8.5**) is based on the work of Joe Leggio, WB2HOL, who applied all the points made above to his original design. The design has been adapted by David Deane, G3ZOI, and this uses UK-sourced plastic components. The dimensions are very close to those originally proposed by WB2HOL.

The original design has been adapted to optimise performance at the IARU-recommended frequency band for ARDF (144.5 - 144.8MHz). The antenna has been modelled using *EZNEC* and the use of a non-conducting boom immediately removes all those compensations that have to be applied to account for the method of mounting the elements on a conducting boom. Fig 8.5 shows the dimensions.

The tape elements were modelled as wire rectangles made of 1mm diameter wire. Hence, the director is modelled with two wires of length 900mm and two wires of length 19mm (for 0.75in-width steel tape), joined to make a rectangle. The wire diameter was chosen to simulate the profile of the curved tape when viewed along the boom. The spacings given in Fig 8.5 are centre-to-centre dimensions for the wire rectangles representing the three elements.

The results of modelling have to be viewed with some caution. Rejection ratios in excess of 30dB require such close amplitude and phase accuracy that it becomes very very hard to make the design reproducible. Just try altering the diameter of cylindrical elements (when used) to see the devastating effect this can have on a finely-honed design. Indeed, the assumptions made in defining the model now become critical. Aim for 20 to 25dB of suppression and the likelihood is that the practical results will follow the model. This model gave 32dB front-to-back and 25dB front-to-side at 145.5MHz (shown in **Fig 8.6**) and the results alter very little as the frequency is changed.

These three-element antennas are widely used in the UK and have always been found to be excellent for ARDF purposes. The excellent side- and rear-rejection predicted by the model is experienced in practice.

The original WB2HOL design is unbeatable for simplicity, but the 4-way pipe connectors used to support the elements are not generally available in the UK. Most plastic pipe made for piping (eg Speedfit) is too soft to be usable for the boom. PVC electrical conduit is sufficiently rigid, but the diameter is on the small side (20mm). This design is based around the Marley 21.5mm diameter overflow system and utilises the associated two-hole-fixing plastic saddle clip.

To attach the elements to the boom, some of the 21.5mm pipe is cut lengthways into 120° segments. These are a reasonable 'match' to

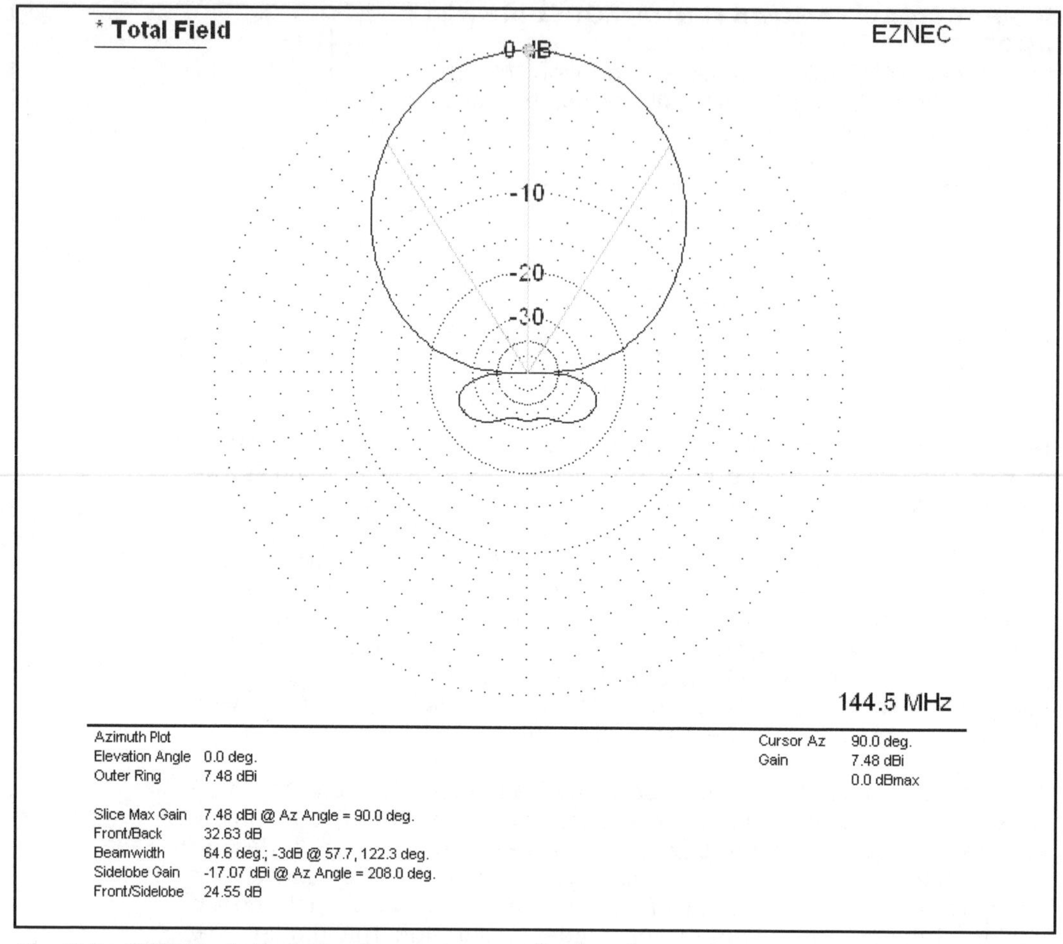


**Fig 8.6: *EZNEC* plot of the free-space radiation from the 144MHz Yagi, using the ARRL-style of amplitude scaling.**

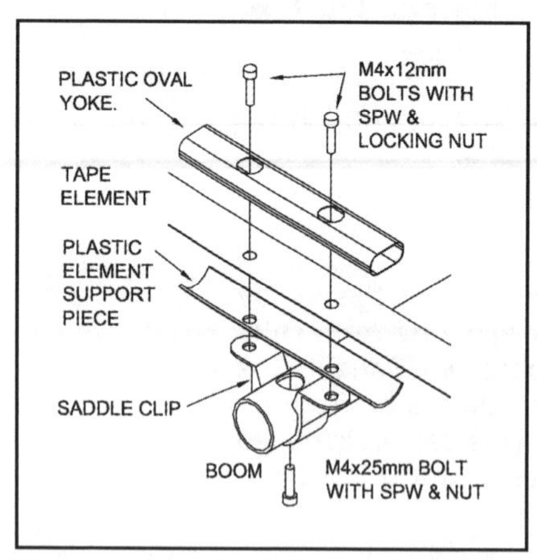


**Fig 8.7: Detail of the mounting of the director and reflector elements.**

the curvature of the steel tape measure. Half-inch wide steel tape should be avoided; it is simply not sufficiently rigid for this application. Tapes are available in 0.75in and 1in widths and both of these are suitable.

The support segments of pipe are 100mm long and are drilled with 4mm holes at 42mm spacing to match the Marley pipe clips. The steel tape is then 'sandwiched' between the support segment and a 100mm length of small oval plastic electrical conduit. The details of the mount are shown in **Fig 8.7**.

The driven element is mounted in exactly the same way, but with a 10mm gap between the two halves of the element. A short length of larger oval plastic conduit sleeves over both the tape

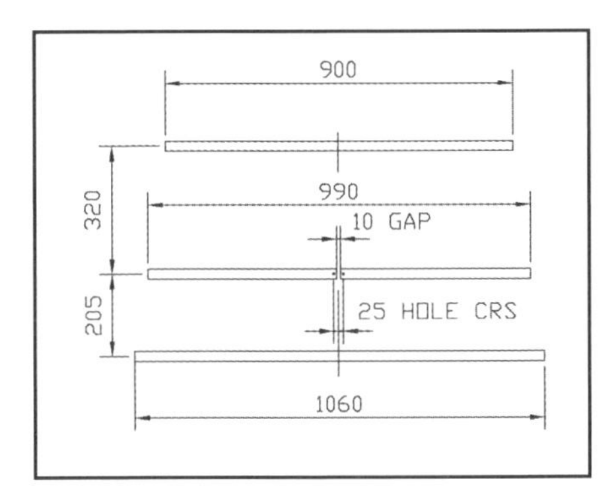


**Fig 8.8: Dimensions of the antenna.**

element and supporting centre plastic section. When used in conjunction with the single M3 bolt in each half, the elements are securely held in the required position. The feeder has to be connected to the element halves and it is possible to solder to the steel tape. The tape should be scraped clean of paint and tinned prior to being placed in contact with the plastic components. In fact, by assembling, then dismantling,

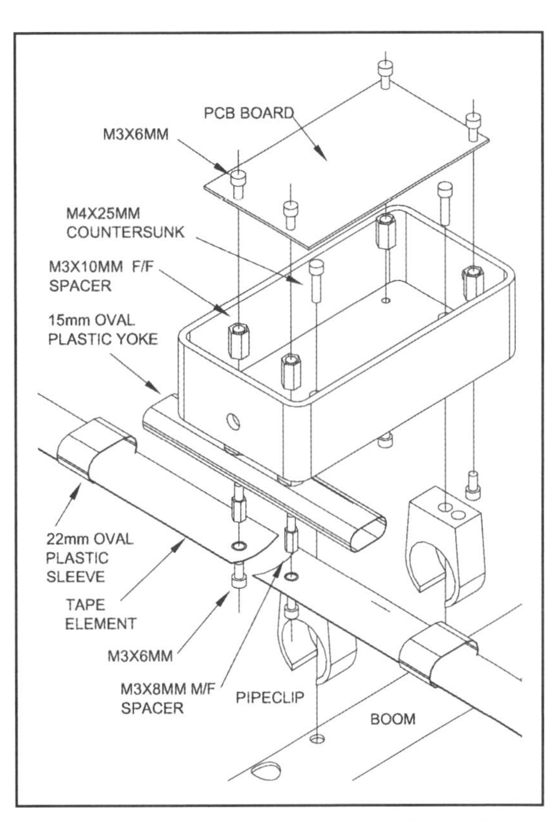


**Fig 8.9: Diagram of the centre of the driven element.**

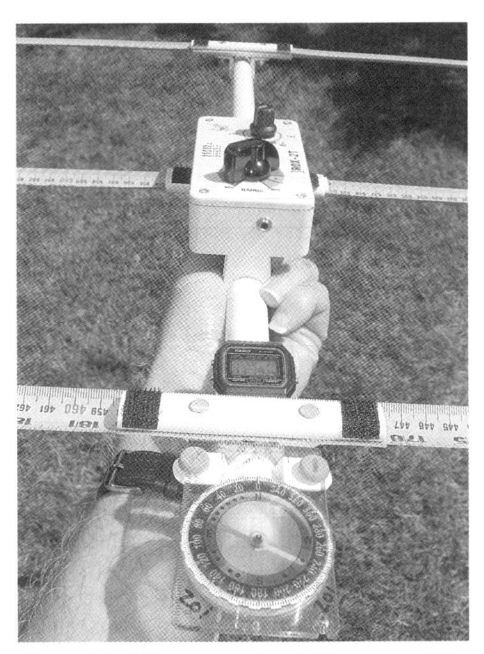

**Fig 8.10: Photograph of the complete antenna showing a ROX-2T receiver, a compass and a watch, mounted on the boom.**

and only then soldering the driven element to the feeder, it is possible to make good electrical connections without melting all the plastic components in the vicinity. **Figs 8.8, 8.9 and 8.10** illustrate the design.

## RECEIVER FOR 3.5MHz

### Direction sensing on 3.5MHz

Directional Yagi antennas are completely impractical on 3.5MHz since one would be 40m wide. ARDF receivers for 3.5MHz use either a ferrite rod or a loop antenna. These couple to the magnetic component of the incoming radio signal. **Fig 8.11** illustrates this. Since the transmissions are vertically polarised, the electric field component (E-field)

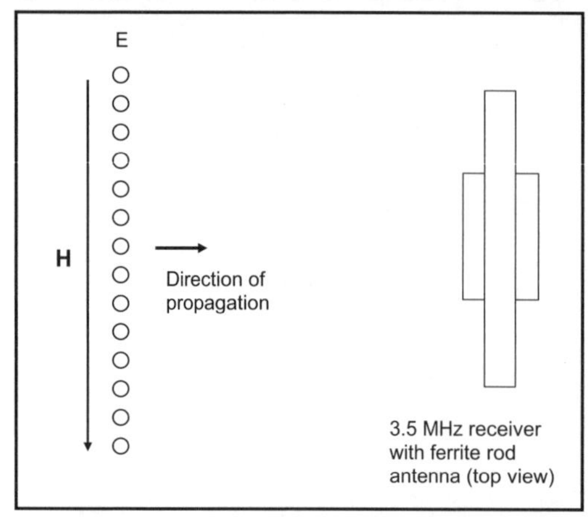


**Fig 8.11: A vertically-polarised electromagnetic field propagating from left to right. The view is from above, looking towards the ground. The direction of the E-field is upwards towards the observer. The ferrite rod in the receiver is aligned with the H-field component to give maximum received signal.**

will be vertical and the magnetic field component (H-field) will be horizontal.

When the ferrite rod is aligned at right angles to the direction of the source transmitter, the received signal is maximum. If the ferrite rod is now rotated by 90° so that it points at the transmitter, as shown in **Fig 8.12**, the ferrite rod is at right angles to the H-field and a minimum signal is received because the lines of force are no longer able to link efficiently with the coil on the ferrite rod. In practice, this minimum is very sharp, allowing accurate determination of the direction to the transmitter.

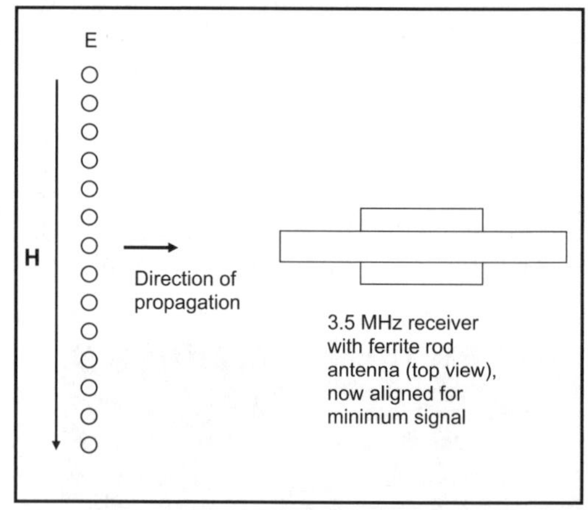


**Fig 8.12: The effect of rotating the receiver by 90°. The ferrite rod now picks up very little energy from the EM wave.**

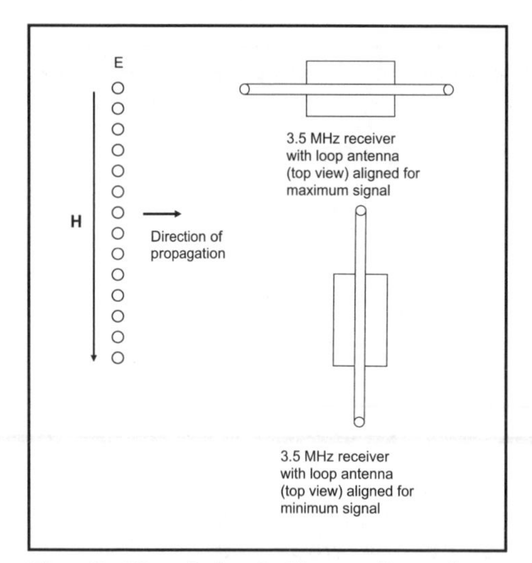


**Fig 8.13: Orientation of a loop antenna for maximum and minimum received signal.**

For receivers using a loop antenna, the plane of the loop points at the transmitter when receiving maximum signal. **Fig 8.13** illustrates this. Loop antennas generally give a sharper null than a ferrite rods but, unlike the ferrite rod, there is no clear 'sighting line' down the line of the antenna towards the transmitter.

Both a ferrite rod antenna and a loop antenna have a 'figure of eight' polar diagram, **Fig 8.14**, and this leads to ambiguity of direction. The nulls are sharp and, as noted above, give a sensitive indication of direction to the transmitter but there are two possible such directions.

If the competitor is standing at the edge of the mapped area, it may be obvious which of the two possible null directions is the correct one.

However, once in the middle of the competition area, this is no longer the case. A means of discriminating between the two directions is required. This is achieved by the use of a second antenna which samples the E-field component of the incoming signal, as shown in **Fig 8.15**. It takes the form of a short whip antenna.

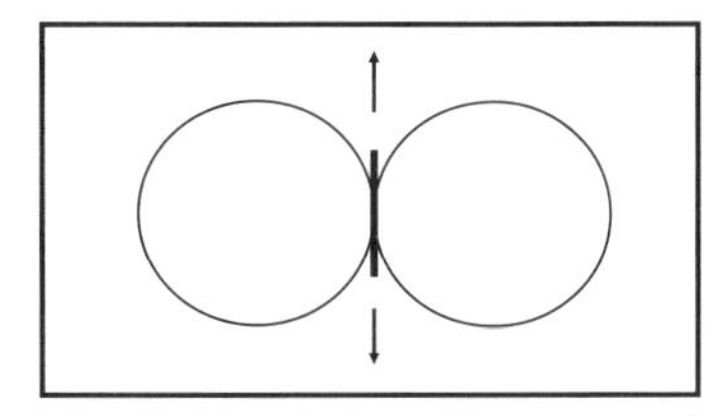

**Fig 8.14: Polar diagram of a ferrite rod antenna. The sharp null, denoted by the arrows, is in line with the ferrite rod shown at the centre.**

The output of the whip antenna is added to the output of the ferrite rod or loop. The polar diagram is modified so that it is now asymmetric as shown in **Fig 8.16**.
It is now possible to differentiate between the wanted and unwanted nulls of the response obtained with the ferrite rod alone. To exploit the above effect, the receiver is rotated by 90° so that the ferrite rod is at right angles to the direction of the nulls. **Fig 8.17** shows how the major lobe of the combined response now points upwards.

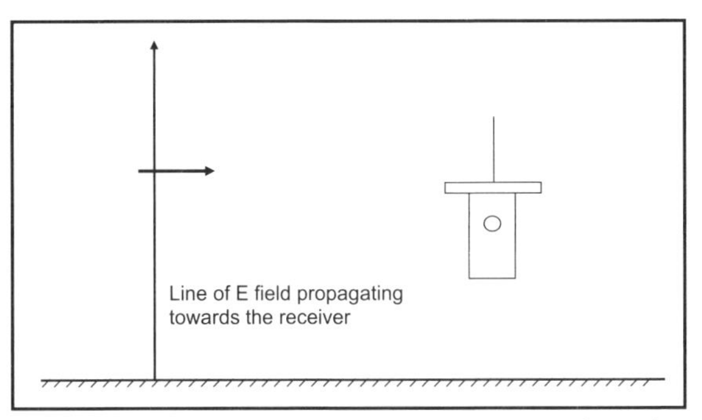


**Fig 8.15: A short whip antenna samples the incoming E-field component which is vertical.**

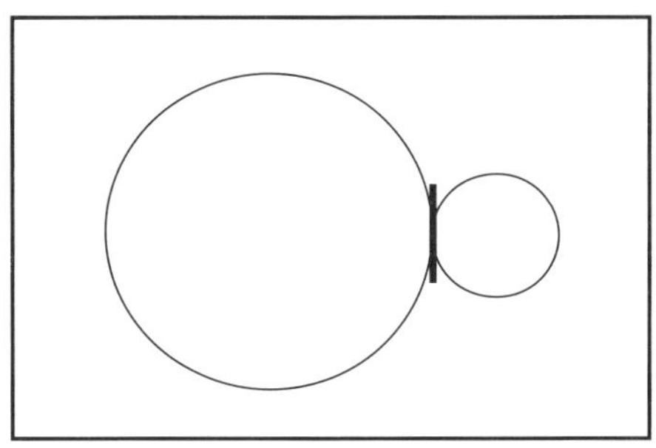

**Fig 8.16: An asymmetric polar diagram results if the output of the ferrite rod or loop is added to the output of the short whip antenna. The position of the ferrite rod is shown at the centre of this polar diagram.**

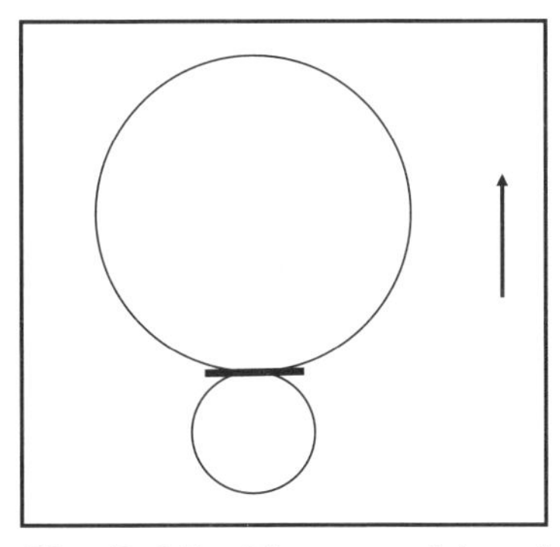

**Fig 8.17: The combined response of the ferrite rod and the whip pointing upwards. The direction of the transmitter is shown by the arrow.**

If the receiver is now rotated by a further 180°, the response points downwards (**Fig 8.18**).

As a result of testing this effect on a transmitter with a known location, the operator will know which of the two nulls obtained with the ferrite rod on its own, points at the transmitter. Hence the ambiguity is resolved. Competitors generally mark their receivers with

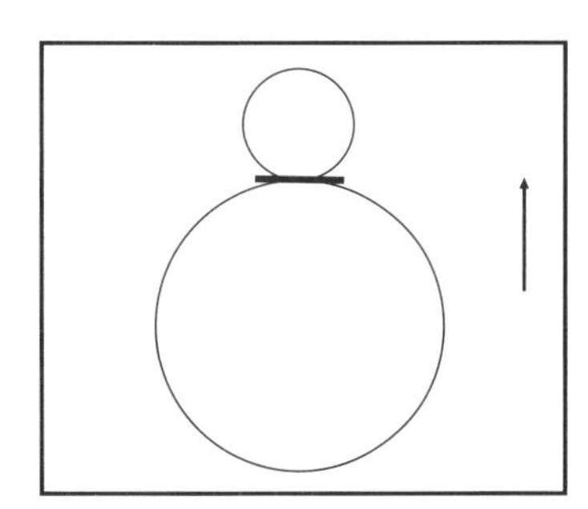

**Fig 8.18: The maximum of the combined response now points downwards and the signal received in reduced, because this is away from the direction of the transmitter shown by the arrow.**

the direction of the transmitter when receiving the stronger of the two signals.

It is important to get an adequate discrimination between the two responses. As little as 6dB is sufficient, especially if the receiver gain is reduced so that the ear is in its most sensitive range for detecting amplitude differences. If the output of the whip is to be added directly to the output of a link coil on the ferrite rod, quite a long whip is required, to the extent that it is unwieldy. Buffering the whip with an FET allows a much-reduced length to be used.

In practice, there is little point in obtaining a very good null in the unwanted direction, for the simple reason that the front-to-back ratio is dependent on the ground conductivity. To understand why this is so, consider the E-field component of the signal arriving at the receiver, having been propagated over poorly-conducting ground. The boundary conditions for an E-field propagating over perfectly-conducting ground are only satisfied if the E-field is at right angles to the conducting surface where they meet. If there were a horizontal component at this point, the perfectly-conducting surface would simply short out the component and eliminate it.

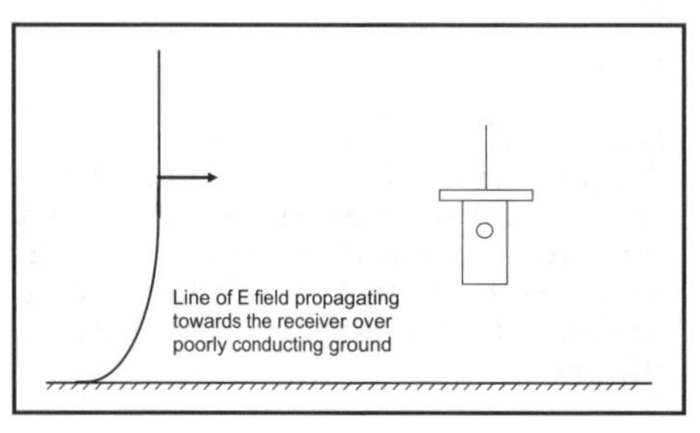


**Fig 8.19: E-field line over imperfectly-conducting ground.**

Over imperfectly-conducting ground, the wavefront is not exactly at right angles to the ground because energy is lost from the wavefront in these circumstances (see **Fig 8.19**). A horizontal component causes currents to be induced in the poorly-conducting ground which leads to energy loss by heating and the additional attenuation of the signal that is observed under these conditions.

If the ground has very poor conductivity the wave front bends even more (**Fig 8.20**) and a greater loss current is induced in the ground.

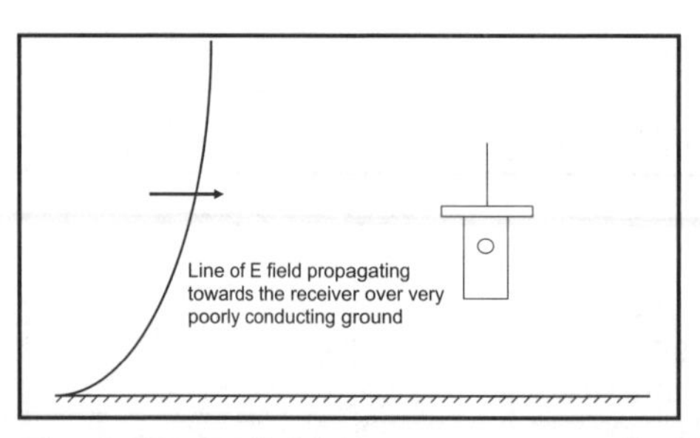


**Fig 8.20: E-field over very-poorly-conducting ground.**

At the height that the receiver is being held in the diagram, the vertical component of the E-field is now reduced due to the bending. If the whip antenna is to pick up the exact amount of signal to cancel the signal delivered by the ferrite rod, the receiver must be held higher in the air in the case of very-poorly-conducting ground. Indeed the height that the receiver must be held is a direct indication of the ground constants (conductivity and permittivity). Hence, there is no point in carefully adjusting the phase and amplitude of the two signals to get a deep null because moving to an area of different ground conductivity will degrade this null making the careful adjustment pointless.

An interesting test is to have the main lobe of the response (whip antenna connected) pointing away from the transmitter and then to move the receiver vertically. If the whip is of the correct length, there will be a point at which the signal is a minimum. Hopefully this is at a height convenient to use in a competition. If not, the whip needs an adjustment to its length.

**3.5MHz RECEIVER**

The receiver described here was originally designed by PA0HRX, and it uses the TCA440 AM radio IC and a low power consumption version of the LM386, to realise a two-chip receiver. The TCA440 is preferred to the TDA1083 used in the simple 144MHz receiver described earlier in this chapter, because it offers a wide gain control range. Varying the DC voltage at the gain control pin controls the gain of three of the four internal IF gain stages. Wide gain control range is a key requirement and the receiver described here has a range of 96dB. The TCA440 is still manufactured in China and it forms the basis of an excellent low-cost DF receiver for 3.5MHz.

This design (**Fig 8.21**)uses a ferrite rod for the main antenna and, to reduce the number of separate windings on the rod, the main winding is capacitively tapped to match into the 4500Ω input resistance of the TCA440.

At least 95% of ARDF competitions using the 3.5MHz band utilise transmitters operating on 3579.5kHz for the simple reason that American colour TV sets using the NTSC system have a colour burst crystal of this frequency. Since they are manufactured in such huge quantities, these crystals are very cheap. The result is that any 3.5MHz ARDF receiver spends most of the time tuned to this frequency with rare excursions to the frequency of the beacon. The tuning range required is 3.5 - 3.6MHz and this covers the frequencies occupied by the foxes and the beacon. Hence the input tuned circuit of this receiver is peaked with a 60pF trimmer at 3579.5kHz.

The sense antenna is a short whip about 150mm in length and this is buffered by a BF245 FET, which has a two-turn link winding, positioned at the centre of the main winding on the ferrite rod, as the drain load. This link serves to add the output of the whip to the signal derived from the ferrite rod before passing it to the input of the TCA440. There is a push-switch in series with the supply to the FET so that the operator can enable the sense circuit at will. The normal procedure is first to locate the sharp null directions without the sense enabled, and only then to press the button and rotate the receiver first by 90° and then by a further 180° to assess which of the bi-directional nulls points at the transmitter.

Selectivity is provided at the output of the mixer and again at the output of the IF amplifier chain by ceramic filters. The filters used have a bandwidth of 6kHz. The external LO components have varicap tuning, and both the coil inductance and the fixed component of the tuning capacitance can be varied to allow good tracking to the frequency scale. Beat-frequency oscillator (BFO) injection from a simple LC oscillator

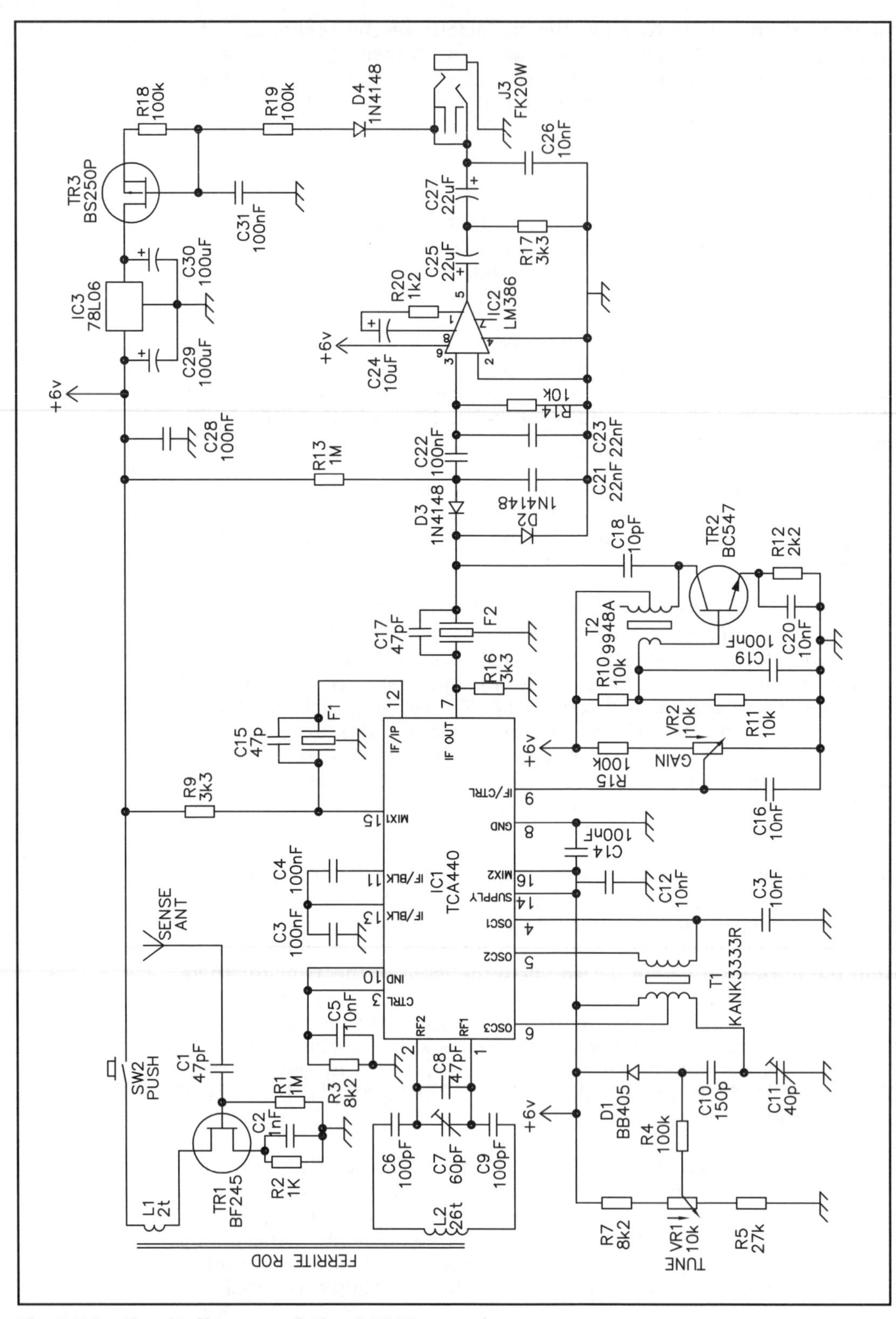


**Fig 8.21: Circuit diagram of the 3.5MHz receiver.**

occurs after the second ceramic filter. The diode detector is forward-biased to improve low-level sensitivity and is followed by an LM386 audio amplifier. The LM386N-1 was used originally, but the version with the lowest power consumption that is available is satisfactory.

The same FET battery switch circuit used in the ROX-2T 144MHz receiver is used here and plugging in the headphones causes a tiny gate current to flow through the headphones turning on TR3, thus connecting the battery to the 6V regulator.

While designed as a low-cost 3.5MHz receiver, the set performs surprisingly well and is capable of good reception of the CW section of the 80m band. After dark, there are many signals to be heard here. The photographs show the receiver in a Hammond case and the populated PCB which fits inside it.

**Populated PCB for the 3.5MHz receiver.**

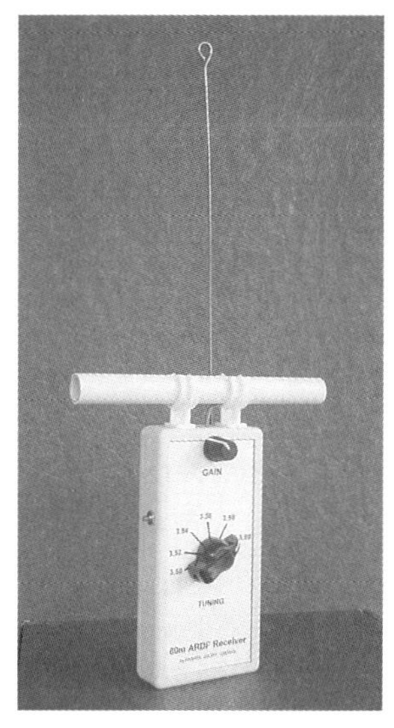

**External view of the 3.5MHz receiver.**

## TRANSMITTERS AND TRANSMITTING AERIALS

### A 1W TRANSMITTER FOR 144MHz

This design (**Fig 8.22**) has been developed by David Deane, G3ZOI, from work done by VE2EMM (the RF strip), ON7YD (the PIC controller) and DL3BBX (the AM modulator). The transmitter is designed for use as the hidden transmitter in direction-finding competitions under the IARU rules. It provides the MCW tone-keyed, amplitude-modulated carrier required by the IARU and, in addition, there is a facility to transmit the frequency-modulated signal suitable for club-based DF hunts, where all the competitors are using FM hand-held radios.

The first thing to strike the reader about the circuit diagram is the apparent simplicity of the RF chain compared with older designs. By using a clock multiplier IC, the only additional active device required is a PA transistor. The clock multiplier is able to deliver sufficient signal power to drive a 2N2247 or a BLT50 to an output in the range 300mW to 1W.

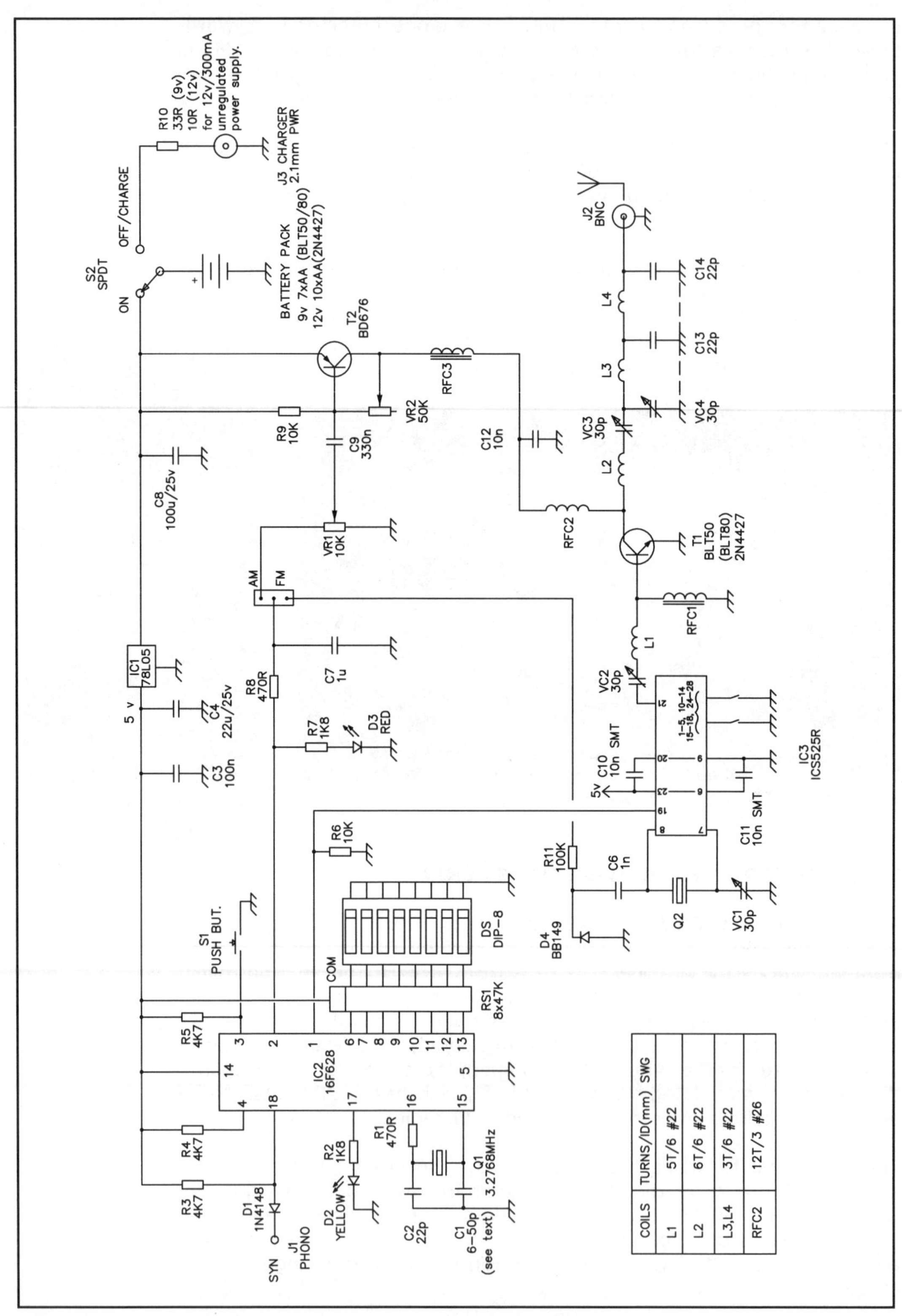


| COILS | TURNS/ID(mm) SWG |
|---|---|
| L1 | 5T/6 #22 |
| L2 | 6T/6 #22 |
| L3,L4 | 3T/6 #22 |
| RFC2 | 12T/3 #26 |

**Fig 8.22: Circuit diagram of the TRO-2 ARDF transmitter.**

The ICS525R clock multiplier is basically a frequency synthesiser designed to provide a clock output determined by the frequency of the crystal used, and user-programmable internal dividers. The internal voltage-controlled oscillator (VCO) can be divided by a three-bit number to generate the output signal, the VCO can be divided by a nine-bit number to provide the signal applied to the phase comparator and the crystal reference can be divided by a seven-bit number to provide the other input to the phase comparator. The number used by each divider is user-programmed by grounding the appropriate pins of the 28-pin SSOP surface-mount package. This gives considerable flexibility in the determination of the output frequency and, in this design, several standard crystals can be used to give an output sufficiently close to the desired frequency that they can be pulled on frequency by adjustment of VC1 in Fig 8.22.

To minimise phase noise from the ICS525R it is a definite advantage to select a crystal requiring the lowest divider ratios to obtain the wanted output frequency. Due to variations in the parallel capacitance assumed by the crystal manufacturer, the crystal may not oscillate exactly on the marked frequency and VC1 on the circuit diagram can be used to give a small adjustment. The 'pulling range' is usually 0 to 30kHz higher than the marked frequency. For example: good results have been obtained using 10.24MHz crystals, programmed with settings derived from the ICS website for 144.500MHz. It is quite feasible to 'pull' the crystal sufficiently to obtain an output on 144.525MHz – the frequency widely used for the hidden transmitters in the UK.

**Suggested standard crystals:**
144.525MHz output (hidden transmitters)

- 14.746MHz (144.507MHz nominal output )
- 10.240MHz (144.498MHz nominal output)

144.775MHz output (beacon)

- 14.318MHz (144.773MHz nominal output)
- 18.192MHz (144.725MHz nominal output)
- 12.288MHz (144.725MHz nominal output)

By way of example, an ICS525 can be programmed to give 144.500 – 144.525MHz from a 10.24MHz crystal. In the VCO divider word, the unwanted division numbers must be grounded, in this case V3, V7 and V8, thus dividing by the sum of the remaining numbers V1, V2, V4, V5 and V6 which is 119. In the Reference divider, R3, R4, R5 and R6 must be grounded to divide by 7. The output divider is set to unity when S1 and S2 are grounded. The output frequency is given by the equation

$$\text{Output frequency} = \frac{\text{Input freq} \times 2 \times (\text{VCO division} + 8)}{(\text{Reference division} + 2) \times \text{Output division}}.$$

The disadvantage of using the clock multiplier IC is that the output is harmonic-rich and needs to be filtered to reduce the level of the unwanted components. The low output of the transmitter does mean that rather less filtering is needed to get these unwanted components down to an acceptable level. A series-tuned circuit is used between the ICS525 and the PA transistor. The latter is operated with zero bias and the output is extensively filtered.

The design allows either the 2N2247 or the surface-mount BLT50 (or BLT80) to be used in the PA. Generic 2N4427s will typically give the minimum specified gain of 10dB, so at least 300mW can be expected. Genuine Motorola 2N4427s typically give 12dB gain and the output will be nearer 500mW. The BLT50 is a 1W / 10dB gain device, intended for use at 470MHz. Down at 144MHz, the gain is considerably greater than 10dB and 1W output is easily obtained when the transmitter is operated in the AM mode. The BLT50 also requires a lower supply voltage. Using a battery pack of seven AA cells and a dummy cell in an eight-cell holder gives a supply of 9V, and this limits the RF output. Although eight cells can be used, most of the power from the additional cell is likely to be dissipated by the BD686 modulation transistor, which will then get quite hot. The current drawn from the supply will be higher with the BLT50 (approximately 270mA) but with a 1:4 transmit : receive cycle and modern AA size NiMH batteries with a capacity of 1500mAh or more, the battery pack will operate for many hours.

**User-programming of the ICS525 by the use of solder bridges to ground the appropriate pins. On the left (top to bottom), bridges 1, 3, 4, 6, 7 and 9 and, on the right (top to bottom), bridges 2, 3, 4, 5, 8, 9 and 10 are made which allows a 14.7456MHz crystal to generate an output on 144.500MHz. Note that these numbers are NOT the IC pin numbers. The ungrounded pins are pulled high internally.**

Amplitude modulation is applied by a BD686 Darlington transistor connected in series with the supply to the PA. 100% amplitude modulation is obtainable with the BLT50 and less downward modulation is experienced than with the 2N4427. Unlike the original DL3BBX design, which used a 555 timer chip to generate the modulating tone, improved code and an upgraded PIC micro-controller IC, generates the keyed tone directly in the PIC.

The PIC chip has been upgraded to a 16F628 which, as well as generating the keyed tone directly, also allows the callsign of the operator to be programmed into the chip by use of a tactile push-button on the PCB. In both the US and the UK, it is a condition of the licence that the callsign of the operator must be sent with each transmission. This is usually done at the end of the one-minute transmission in a burst of higher speed CW. To provide the extra software driven-functionality, the original ON7YD assembly code software used to program the PIC has been replaced with new code written in 'C', derived from code published by VE2EMM. The operational format of the transmitters, however, remains largely compatible with the ON7YD designs.

 *Hex code for the above can be found at http://www.rsgb.org/books/extra/ardf.htm*

The PIC is also provided with a phono socket connected to pin 18 for synchronisation purposes. A set of DIP switches is used to tell the transmitter which one it is in the sequence. The first transmitter will radiate MOE for its identification. Switches 1 – 3 are used for this purpose, giving eight possible combinations: five for hidden transmitters; one for the beacon; and the remaining two are often used to provide for a two-transmitter two-minute cycle DF hunt. The remaining five switches are used to set time delay and the standard arrangement is to have additive delays of 0.5, 1, 2, 4 and 8 hours. The transmitters to be hidden are brought together in one place prior to the start time, all the sync ports are connected to a common switch and at the desired instant the switch is opened to tell each transmitter that it is time 'zero'. If the transmitters were required to 'fire up' starting at 1030, then synchronisation at 0730 would require a three-hour delay to be programmed into the DIP switches.

Transmitter one would then come on the air after three hours exactly and start sending MOE for one minute. Transmitter two would come on after three hours and one minute and start sending MOI for one minute, transmitter three would come on after three hours and two minutes and start sending MOS and so on...

Please note : Attention to the PIC clock frequency is essential if the units are to be used in serious events, or if long delay times are used; otherwise, cycle times and synchronisation will drift, the latter resulting in an annoying crossover period with heterodynes, between two transmitters which are adjacent in the sequence.

Replacing C2 with a variable capacitor (5 – 50pF) will usually enable the crystal frequency to be pulled to the fundamental frequency of 3.2768MHz. More important is to zero-beat each of the timer clocks in the transmitter set with a master transmitter. This will ensure synchronisation is maintained over very long operational periods exceeding 24 hours. Constructional details can be found in **Figs 8.23** and **8.24**.

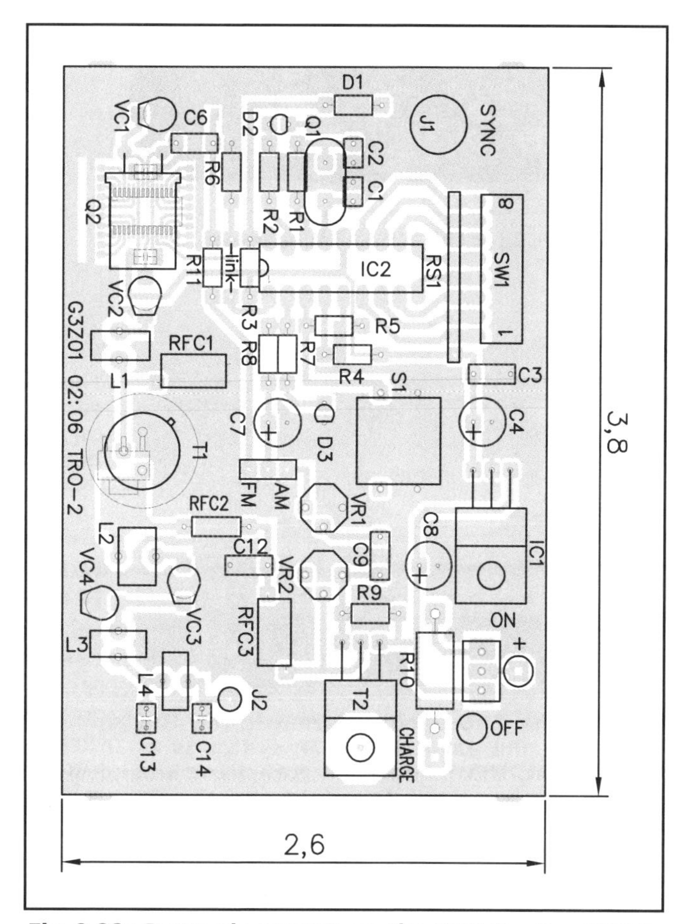


Fig 8.23: Parts placement on the PCB for the TRO-2.

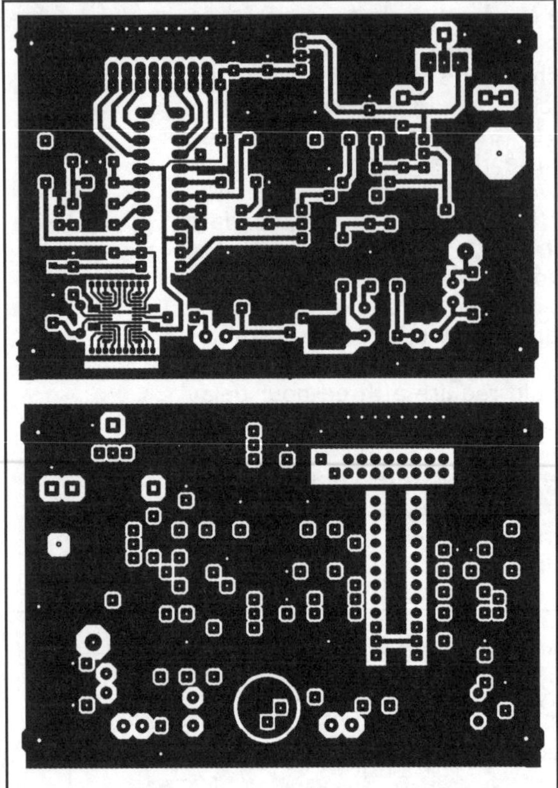

Fig 8.24: PCB for the 1W 144MHz transmitter.

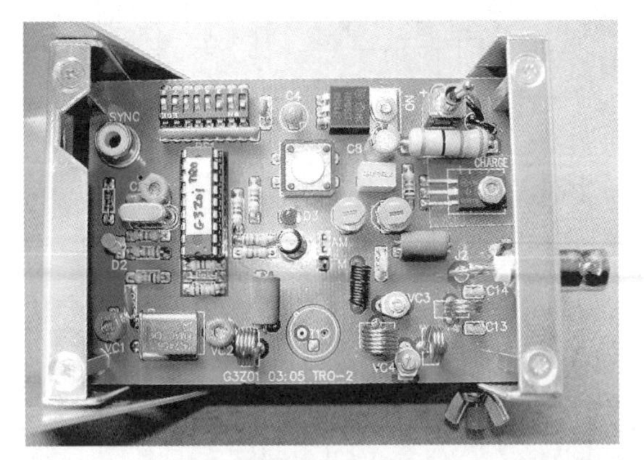


View of the TRO-2 circuit board. The PIC controller and associated DIP switch is at the top left, the modulation transistor and voltage regulator are top right and the RF strip and PA are across the bottom. The push switch near the top centre is used when programming a callsign chosen by the user, into the controller.

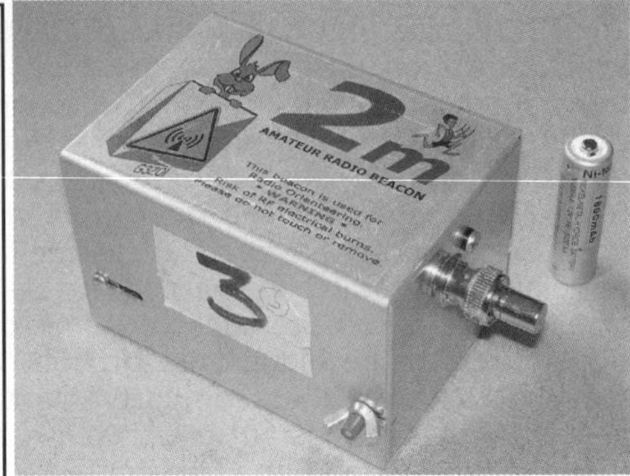


View of the TRO-2 in its case.

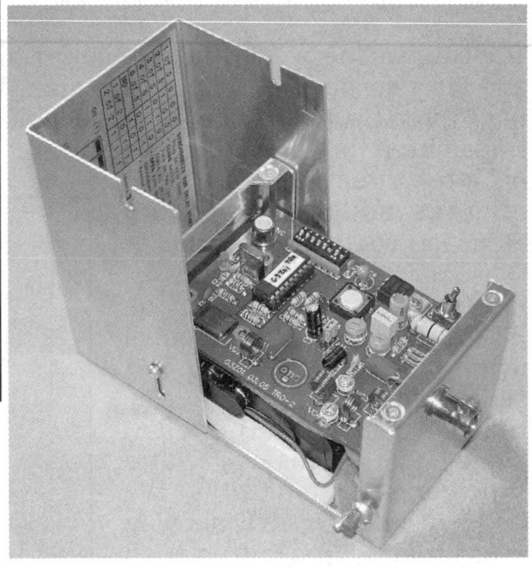

The TRO-2 with the lid lifted. DIP switch settings and operating instructions are on the inside of the lid. This version has the surface-mount BLT50 under the PCB as is the ICS525 clock multiplier chip.

## TURNSTILE ANTENNA FOR 144MHz TRANSMITTER

This is the most widely-used antenna for 144MHz ARDF transmitters. The requirement is for a horizontally-polarised, omnidirectional antenna. The turnstile design is straight out of the RSGB *VHF/UHF Handbook* of the 1970s (**Fig 8.25**). It comprises a pair of crossed dipoles with a quarter-wave of 75Ω coax linking the dipoles and the feeder connected to one of the dipoles.

It is essential that the elements can be removed from the rest of the antenna assembly so that the antenna can be transported easily. In the design presented here, the elements are made of 18SWG brazing rod, soldered into 4mm wander plugs. The centre junction of the antenna uses a small square plastic box with a short 4mm socket mounted on each face. This arrangement makes it easy to get the four elements mutually at right angles to each other. A suspension eye is mounted in the centre of the top of the box, and this can be made from a vine suspension eye with the woodscrew part cut off and a suitable machine thread tapped onto the shank. (In the UK, vine eyes are available from B & Q and Screwfix,) Three holes are drilled in the bottom lid to accept the 75Ω phasing coax (two holes) and the 50Ω feeder respectively. Assembly of the antenna takes literally seconds to push the elements into the sockets. The springs on the plugs hold the elements firmly in place. The photographs illustrate the assembly.

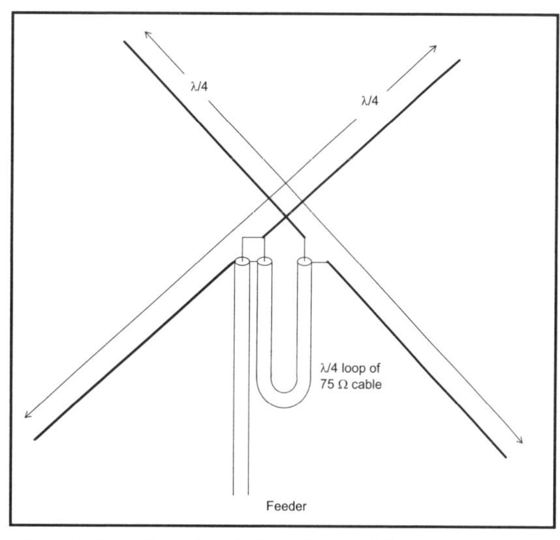


**Fig 8.25: Circuit of the turnstile antenna. It consists of two crossed half-wave dipoles. One dipole is fed directly from the feed-line, while the other is connected by a quarter-wave phasing line made from 75Ω coaxial cable.**

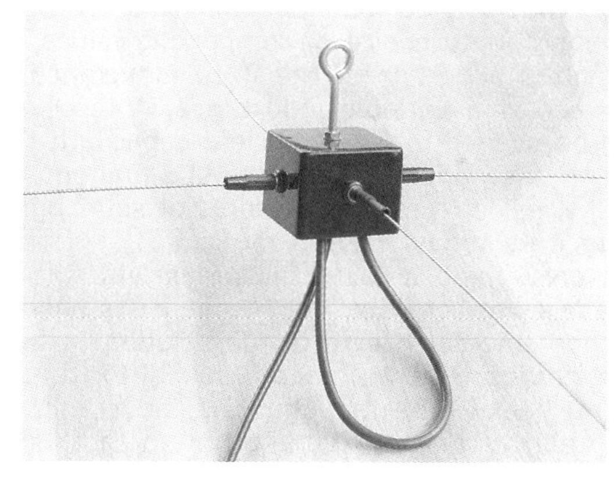

**A turnstile antenna for 144MHz transmitters. The quarter-wave elements are fitted with 4mm wander plugs and are fitted into sockets on the four faces of a small plastic box. The suspension eye is seen at the top and the quarter-wavelength loop of 75Ω coaxial cable below the box is the phasing line between the crossed dipoles.**

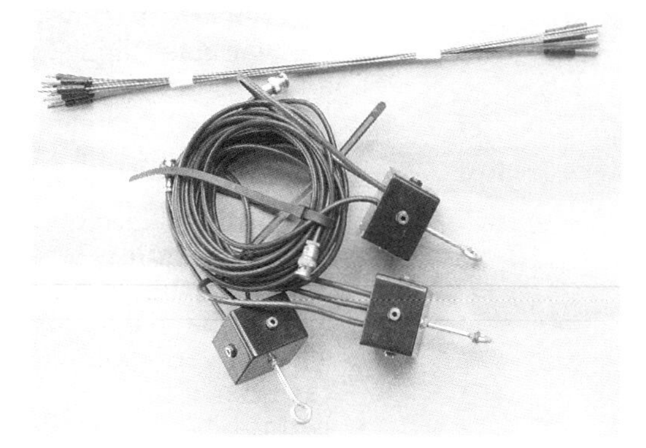

**Three turnstile antennas for 144MHz, packed ready for deployment.**

To reduce the risk of injury from the ends of the elements (which are quite sharp) protective caps can be made by drilling push-fit holes in short lengths of dowel.

## ANTENNAS AND TRANSMITTERS FOR 3.5MHz

Before building a transmitter for 3.5MHz, an important decision has to made with regard to the antenna. Resonant antennas on 3.5MHz are impractically long (a quarter-wave is 66ft / 20m). Therefore some kind of tuning or loading of the antenna has to be included.

There are two approaches.

- A wire antenna, significantly shorter than a quarter-wave, which will require tuning to resonance.
- A short self-resonant antenna, which is tuned by virtue of having either a loading coil or a helical winding on a former. This is referred to as a loaded antenna.

The former approach is encountered more frequently and it is usual to use an 8m wire with an 8m counterpoise. This wire length (26.4ft) is about the longest length that is easily supported by throwing a line over a tree branch and pulling up the antenna. A single counterpoise reduces the ground losses to some extent and it is usually positioned so as not to trip competitors arriving at the transmitter. The antenna and the counterpoise are just pieces of wire and are light and easily carried.

The 8m wire will need to be tuned to resonance, and there are three ways in which this can be achieved in a compact format.

- Tune the antenna system with a fixed inductance in series with a variable capacitor. By appropriate choice of component values, the variable capacitor cancels a portion of the fixed inductance and the net effect is to act as a variable inductance. A three-gang variable is likely to be needed to give sufficient capacitance swing to tune the wires to resonance in a variety of situations. This solution is quite bulky, aside from the problems of sourcing three-gang variable capacitors today.
- Switch a series of taps on a fixed inductor in series with the antenna. This can be made quite compact, but resonance is only approximate due to the incremental nature of the solution.
- Make a variometer to do the job. The variometer is ideal for this task since its normal drawback – the fact that the inductance cannot be reduced to near zero – is not a problem in this application.

The loaded antenna approach has been used by Rik Strobbe, ON7YD, and David Deane, G3ZOI. G3ZOI uses a 6m fibreglass roach pole with a loading coil wound about a metre up from the base. About 70 turns of 26SWG wire are used for the coil and a 5m wire is attached to the pole above the coil. By winding the coil at the top of the bottom section of the roach pole, the pole can still be telescoped easily for transportation.

The pole is supported on a 10mm-diameter copper-covered steel ground rod. This is hammered partially into the ground to provide the earth connection and the part that protrudes is sleeved inside the roach pole and keeps this vertical. **Fig 8.26** shows the concept.

The advantages and disadvantages of the two approaches are:

- *Ease of erection.* The roach pole vertical can be erected easily in any location, whereas the 8m wire needs a reasonably-sized tree to support it. It takes a lot longer to get a supporting string for a wire antenna over a branch 7m or 8m above ground than it does to erect a roach pole and hammer in the supporting ground spike.
- *Concealment.* The roach pole antenna is much more visible than a piece of brown or olive green wire and potentially might be liable to theft. Competitors on occasions locate the transmitter by seeing the pole before seeing the low hung banner.
- *Ease of deployment.* The roach pole vertical requires the pole and a ground rod plus a big hammer! If more than two transmitters are to be deployed on a single walk through the terrain then careful consideration has to be given to the problem of carrying the equipment. It is much easier to carry five 8m wires and five 8m counterpoises than it is to lug five roach poles and five ground rods.
- *Transmitters.* The transmitters used with a tuned antenna can have a nominal 50Ω output whereas those for use with the 8m wire require an internal antenna-tuning facility. Hence a decision is needed on the type of antenna to be used prior to building the transmitter. If the transmitter has an internal tuning arrangement for an 8m wire, it will also need an manual PTT switch to operate the transmitter for tune-up purposes. The antenna tuning facility clearly adds to the cost of building a set of six transmitters.
- *De-tuning.* The fixed tuning of loaded antennas is prone to detuning when placed in or near wet vegetation.

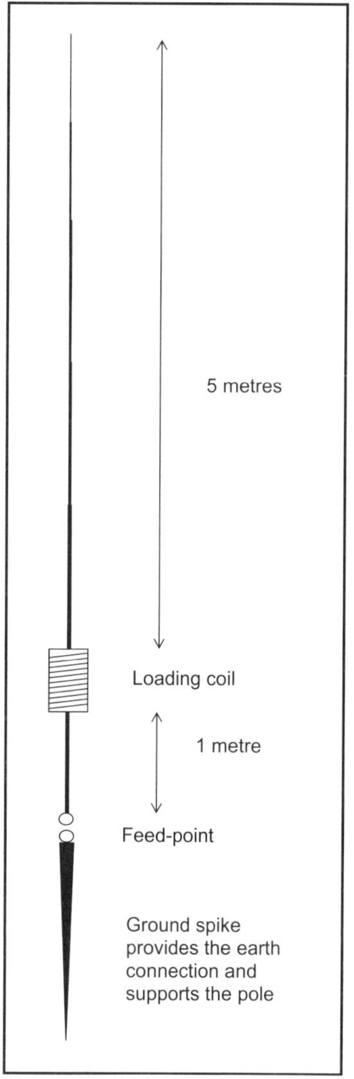


**Fig 8.26:The G3ZOI loaded vertical antenna for 3.5MHz.**

Finally, a note about antenna efficiency. The radiation resistance of an 8m wire on 3.5MHz is just over 4Ω. The measured feed-point resistance at resonance of either type of antenna is likely to be in excess of 100Ω. Hence antenna efficiencies of 4% or less are to be expected. This sounds appalling, but the practical experience is that a 3W transmitter radiating about 120mW from this 4% efficient antenna, is audible over a distance of several miles - sufficient for a DF hunt over the largest areas used.

The transmitter to be described later delivers essentially the same power into a range of load resistances. It is therefore only necessary to tune the antenna to resonance. Matching of the interface between transmitter and antenna to exactly 50Ω is not required. Hence, tuning the antenna is a simple single-knob operation to get maximum output.

## ANTENNA TUNING

### Switched inductance

This is a straightforward method of adding sufficient series inductance to the antenna to bring it to resonance.

A tapped coil using a single-pole 12-way rotary switch to select the appropriate coil tap is a straightforward way of implementing this idea. A iron dust core is far more suitable for this application than ferrite, and experiments have shown that a ferrite core of a similar size can give 6dB additional loss compared to iron dust. The problem with iron dust cores is that many turns are required to achieve the inductances required to resonate an 8m wire on 3.5MHz.

A T94-2 iron dust core (colour coded red for all cores made of #2 mix; the '94' refers to the core outer diameter of 0.94in) is a good choice and this should be wound with enough turns to achieve an inductance of 20µH. This will be about 40 turns depending on the permeability of the individual core and also on the way in which the turns are spread out around the core. Use wire of about 0.2mm OD and wind close-spaced. If at all possible, check that an inductance of 20µH has been achieved and adjust the number of turns as necessary.

**A switched inductance tuner. A single-pole 12-way rotary switch is used with a T94-2 iron dust toroid and the tuner pictured gives 12 values of inductance between 20µH and 48µH.**

Next, wind a further 22 turns with 11 further taps having two turns between each tap. (The first tap is at the end of the original winding of 20 turns) This will give a total of 62 turns and an inductance of about 48µH. The inductance range of 20 – 48µH is more than sufficient to tune an 8m wire with an 8m counterpoise. This solution is compact and easy to implement. The disadvantage is that, in some situations, tuning exactly to resonance will not be possible and there will be a small reduction in antenna current as a result.

### A variometer

The variometer is an ideal component to provide the variable inductance to tune a short wire antenna to resonance on 3.5MHz. The variometer comprises two coils, one of which is allowed to rotate inside the other. The coils are connected in series and when their

**Variometer coils photographed before assembly. The outer coil is on the right. The inner coil has flexible flying leads to allow is to be connected in series with the outer coil.**

axes are aligned but the magnetic fields oppose, the inductance is a minimum. Rotate one of the coils by 180° and the magnetic fields will aid each other and the inductance is a maximum. The variometer cannot give very low inductances, because this requires the two coils of the same inductance to have a coupling factor of 1, but it readily gives a variable inductance of 15 – 50μH which is quite suitable for tuning an 8m wire on 80m.

The first photograph shows how the outer fixed coil (right hand side) is wound on a short length of 36mm-diameter plastic waste pipe. 30 turns of 0.5mm-diameter enamelled wire are wound with 15 turns on each side of small brass bearings, made by turning down M4 brass bolts and securing them with M4 brass nuts. The inner coil former is a piece of plastic overflow pipe 21mm in diameter. 10+10 turns of the same 0.5mm-diameter wire are wound here. A shaft is made of M3 studding to facilitate the locking of the components into position allows the smaller coil to be placed inside the larger one and to be rotated. A purpose-made bushing and 0.25in shaft incorporating the end stops completes the variometer.

The second photograph shows the variometer installed in a complete 80m ARDF transmitter. The ON7YD transmitter with a PIC timer providing the keying, timing and synchronisation required, is on the left. The battery is a lead-acid gel-cell at the bottom and the variometer is seen on the right. The antenna can be tuned by simply listening on an 80m DF receiver and tuning for loudest signal when the transmitter is operating. A PTT pushbutton is provided on the panel of the transmitter, so that the transmitter can be activated for antenna tuning purposes even when the PIC timer is in time-delay mode.

**A variometer installed alongside an ON7YD transmitter in a diecast box.**

The variometer described here is cheap in terms of the materials needed, but expensive in terms of time. Access to a small lathe is needed to make the small brass component parts. This variometer has an inductance range of 15 – 50μH.

### The ON7YD transmitter

This versatile transmitter uses just a CMOS 4001 quad NOR gate IC and an FET to generate 3W of RF on 3.5MHz. The circuit is shown in **Fig 8.27**.

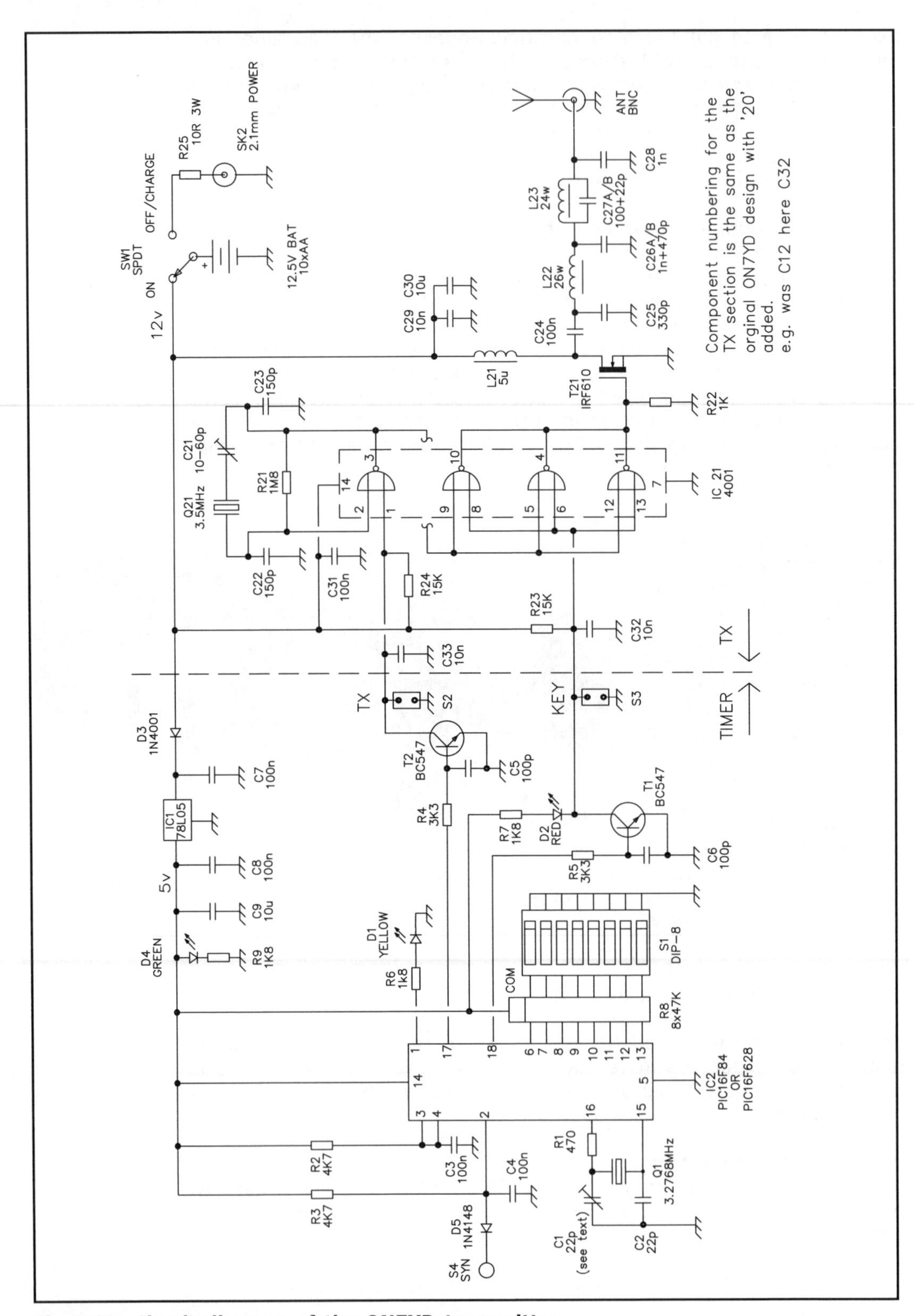


**Fig 8.27: Circuit diagram of the ON7YD transmitter.**

One gate of the 4001 is used as part of a crystal oscillator using a very familiar circuit arrangement. The frequency chosen for the hidden transmitters is nearly always 3.5795MHz, because this is the frequency of the colour burst crystal in a TV set using the NTSC system. Hence the crystals are readily available and very cheap.

The remaining three gates of the 4001 are paralleled to form a buffer amplifier which drives the gate of the IRF610 FET. This mode of operation of CMOS gates does work, but different versions of the 4001 perform differently. For this reason it is a good idea to use a socket for the 4001 and then ICs can be swapped around if there is any shortage of drive to the IRF610. The drain current of the IRF610 is rich in harmonics and a two-stage low-pass filter follows the PA to give a reasonably pure output sine-wave.

The PIC microcontroller shown is a development of the original ON7YD design. The PIC16F628 is a more modern controller than the 16C84 and also has the advantage of being less expensive as well as offering greater internal memory. The callsign of the operator can be included at the end of every transmission, as required by the licensing authorities in some countries. However, using the EEPROM, the call stored can be changed in the field without reprogramming the whole chip. Attaching a switch to the *Sync* port of the PIC (pin 2), in conjunction with the DIP switches, allows the user to program in the callsign to be sent at the end of each transmission. This offers great flexibility and the callsign can be modified when required.

Note that this software for the 16F628 is not the same as that used on the 144MHz transmitter described earlier. This version is fully compatible with the pin-out of the original ON7YD design for the 16C84. The 16F628 offers nearly double the program capacity as the 16C84, and with this version of the software it is only about 50% utilised.

## TRANSMITTERS FOR 3.5MHZ FOXORING

### Simple types

The ON7YD circuit comprises a 4001 CMOS chip using one gate as a crystal oscillator driving a second gate, which is connected to a short antenna via a transformer. See **Fig 8.28**. The remaining two gates

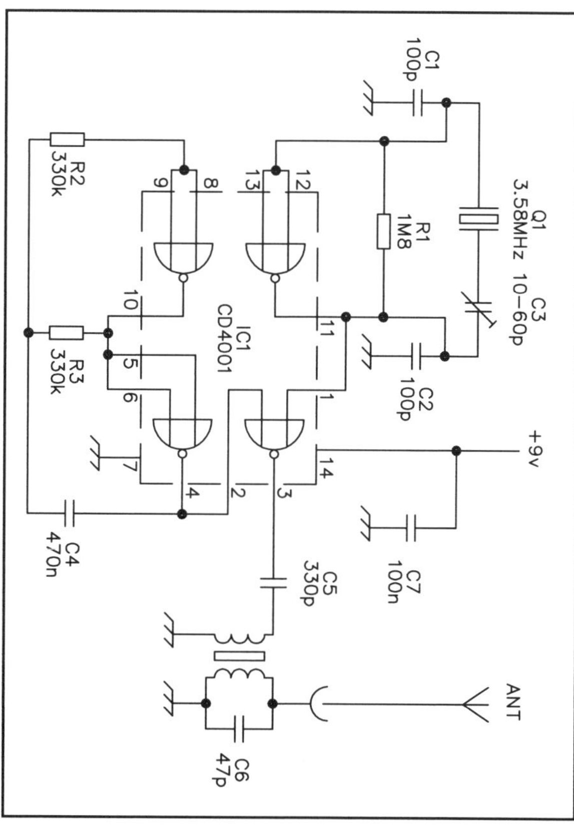


**Fig 8.28: Circuit of the ON7YD very-low power 3.5MHz transmitter.**

are connected as a relaxation oscillator which keys the RF output through the second input to the gate which buffers the crystal oscillator. This oscillator runs at a frequency of less than 1Hz determined by R3 and C4.

The output is coupled to the short wire antenna, which will exhibit a small capacitive reactance in series with a resistance dominated by the earth losses, by a step-up transformer. A ratio of 1:20 was used by ON7YD.

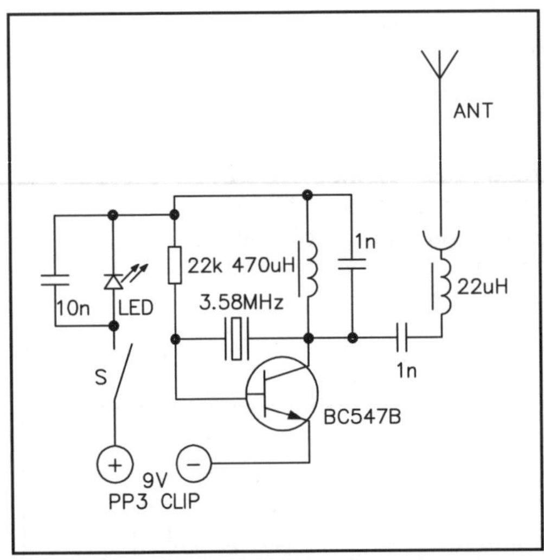


**Fig 8.29: Circuit of the DF7XU very-low power 3.5MHz transmitter.**

In contrast the DF7XU design (shown in **Fig 8.29**) uses a single-transistor crystal oscillator coupled to the antenna via a 22µH inductor which does something to bring the short antenna nearer to resonance. There is an innovative use for a flashing LED in series with the supply, to key the output at a slow rate.

Both of these circuits offer simplicity and low cost. With minimum component solutions like these, transmitters are easily built and the large number needed for a foxOring competition are not a daunting task to build. In some ways they hardly qualify as transmitters. They are designed to be audible between 50 and 100m from their location, the latter distance being achievable with a reasonable superhet receiver. The power radiated is minute and there is precious little difference between a crystal oscillator built on the workbench and one of these 'transmitters'.

Since the keying is just low frequency on / off keying, there is no identification of individual transmitters. Also, they will not comply in countries that require the callsign to be sent by unattended transmitters.

**Sophisticated types**

Whilst still maintaining a large element of simplicity, the next stage up is well-represented by the design shown in **Fig 8.30**. The RF chain owes much to the ON7YD design, except that the two gates, previously configured as a relaxation oscillator, are now paralleled to drive the antenna through a power control potentiometer and a step-up transformer wound on a T50-6 iron dust core. The power control feature is useful if the transmitter is to be used for different roles. For a foxOring competition, the power will be reduced in order to keep the maximum range down to 100m or less but, for a demonstration event, a higher setting will probably be necessary.

The innovative feature of this design is the use of the carrier crystal on 3.57954MHz as the clock crystal for the PIC chip. An output from pin 10

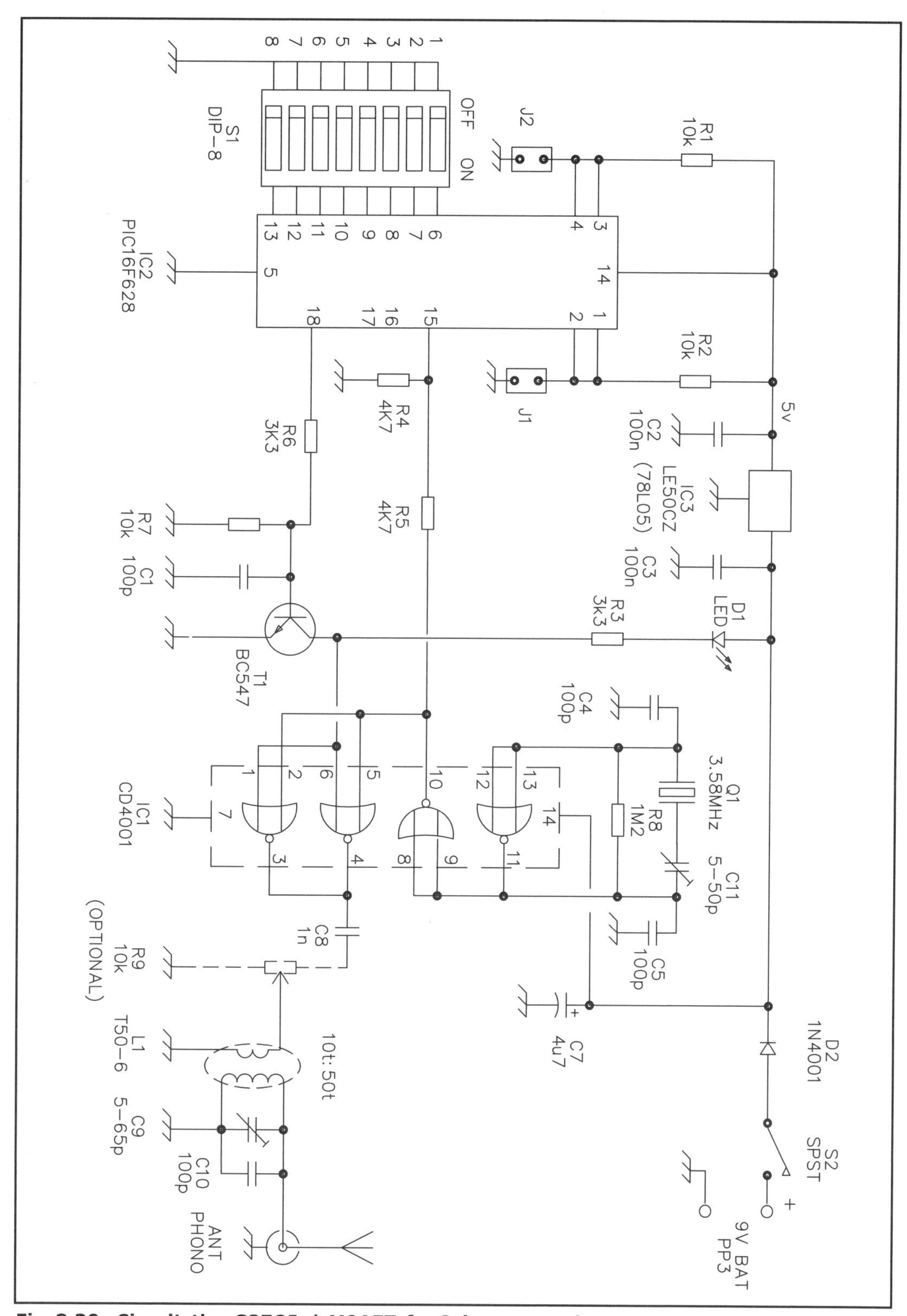


**Fig 8.30: Circuit the G3ZOI / M0AET foxOring transmitter.**

of the 4001 is taken to the clock port of the 16F628. The keying output from the PIC is buffered by a BC547 and drives a LED as well as keying the parallel pair of gates.

With a PIC chip included it is possible to implement a wide range of strategies. Here are a few of the possibilities.

- Include the callsign of the 'operator' at intervals.
- Send a different 'identifying' letter of the alphabet for each transmitter. In a foxOring competition, where 20 or more transmitters may be deployed, this gives the competitor double confirmation of the identity of the station that is located.
- Send a conventional five-transmitter/five-minute cycle using MOE, MOI etc for the identifying modulation.
- Implement a three-transmitter cycle over 90s with each transmitter sending for 30s. The identifications MOE, MOI and MOS can be used. This format is particularly useful for small demonstration DF hunts on school playing fields, exhibition halls and hotel grounds.

The component count has increased significantly compared with the simple types of foxOring transmitters. A PIC, DIP switch, 5V regulator and a transistor are the most significant. The component cost will be greater as a result.

 *Hex code for the above can be found at http://www.rsgb.org/books/extra/ardf.htm*